INTERNATIONAL TECHNOLOGY TRANSFER

INTERNATIONAL TECHNOLOGY TRANSFER:

Concepts, Measures, and Comparisons

Edited by
Nathan Rosenberg
and
Claudio Frischtak

Sponsored by Subcommittee on Science and Technology Indicators, Committee on Social Indicators, Social Science Research Council

PRAEGER SPECIAL STUDIES • PRAEGER SCIENTIFIC

New York • Philadelphia • Eastbourne, UK
Toronto • Hong Kong • Tokyo • Sydney

Library of Congress Cataloging in Publication Data
Main entry under title:

International technology transfer.

Includes bibliographies and index.
1. Technology transfer—Congresses. I. Rosenberg, Nathan, 1927- . II. Frischtak, Claudio.
T174.3.I557 1985 338.9'26 85-6276
ISBN 0-03-000494-2 (alk. paper)

Published in 1985 by Praeger Publishers
CBS Educational and Professional Publishing, a Division of CBS Inc.
521 Fifth Avenue, New York, NY 10175 USA

56789 052 987654321

Printed in the United States of America on acid-free paper

INTERNATIONAL OFFICES

Orders from outside the United States should be sent to the appropriate address listed below. Orders from areas not listed below should be placed through CBS International Publishing, 383 Madison Ave., New York, NY 10175 USA

Australia, New Zealand
Holt Saunders, Pty, Ltd., 9 Waltham St., Artarmon, N.S.W. 2064, Sydney, Australia

Canada
Holt, Rinehart & Winston of Canada, 55 Horner Ave., Toronto, Ontario, Canada M8Z 4X6

Europe, the Middle East, & Africa
Holt Saunders, Ltd., 1 St. Anne's Road, Eastbourne, East Sussex, England BN21 3UN

Japan
Holt Saunders, Ltd., Ichibancho Central Building, 22-1 Ichibancho, 3rd Floor, Chiyodaku, Tokyo, Japan

Hong Kong, Southeast Asia
Holt Saunders Asia, Ltd., 10 Fl, Intercontinental Plaza, 94 Granville Road, Tsim Sha Tsui East, Kowloon, Hong Kong

Manuscript submissions should be sent to the Editorial Director, Praeger Publishers, 521 Fifth Avenue, New York, NY 10175 USA

Foreword

In June 1983 the Social Science Research Council, through its Subcommittee on Science and Technology Indicators, convened a conference to discuss alternative concepts, measures, and comparisons related to the process of the international transfer of technology. The papers presented at that conference constitute the chapters of this volume. The importance of this topic cannot be overstressed. Technical progress and economic growth have historically been closely intertwined phenomena, with both indigenous technological activity and transfers of foreign technology acting as prime levers of economic development and structural change.

In the last two decades much has been written on the nature and impact of the transfer of foreign technology, particularly with respect to developing countries. Some years ago it was believed that the opportunities offered by the availability of more advanced technologies would vastly simplify and accelerate the process of economic growth in developing countries. After all, a peculiarity of invention as an economic activity is that something has to be invented only once. Presumably, after that, it would be merely a matter of acquiring the technology through transfer.

It is now broadly accepted that these sanguine expectations were highly naive. Technology transfer, we are now aware, is a much more complicated process. Not only are such transfers very costly, but their ultimate success is far more contingent upon a number of technical and economic factors than was previously realized. Among the most important of these factors are the level and direction of indigenous technological efforts as well as numerous aspects of the institutional setting in the recipient country. This is made abundantly clear in the chapters of this volume, which discuss the process of technology transfer with reference to the comparative experience of both advanced and developing countries. We find in these chapters a range of common themes and empirical recurrences that shed considerable light on both the possibilities for an effective technology transfer and the features of individual countries that regulate and set the limits for their successful absorption of foreign technologies.

Preface

THE NATURE OF TECHNOLOGY

Received theory conceives technology as information necessary to design and produce a given good by any number of alternative methods. This information is assumed to be public in nature; it is freely available and costlessly reproducible and exists in explicit form and is codified in designs, operating manuals, and so on. Many chapters in this volume argue that, instead of being regarded as public information, technology might be more usefully conceptualized as a quantum of knowledge retained by individual teams of specialized personnel. This knowledge, resulting from their accumulated experience in design, production, and investment activities, is mostly tacit, that is, not made explicit in any collection of blueprints and manuals. It is acquired in problem-solving and trouble-shooting activities within the firm, remaining there in a substantially uncodified state. It is a form of know-how that is open, in the sense of having a nonproprietary character, but this does not imply that other firms would have easy access to it. On the contrary, each individual firm is a locus where the progressive accumulation of technical knowledge takes place, with production processes tending to display many specific and idiosyncratic components. As Richard Nelson aptly put it, in commenting on one of the chapters of this book,

> Firms should not be viewed as entities facing a set of explicit technological possibilities. Rather, they need to be understood as entities which acquire the capabilities they have through a time-consuming and expensive process of learning. This means, at once, that the set of technologies a firm can use effectively at any time is far smaller than the set of technologies that are open to it, and that each firm is relatively unique in the capabilities it has because its learning experiences are to some extent idiosyncratic.

A similar perspective is shared by Pavitt in Chapter 1. Technology, as a highly differentiated range of techniques and of related knowledge, could not "easily be derived or reduced to, first scientific principles; They depend therefore on a range of acquired skills, practices and subtheories (or rules of thumb)." These reflect cumulative processes, where time enters in an essential way. As Baranson and Roark state

> The tens of thousands of elements required for a single industrial product, such as a high-speed diesel engine, are meticulously accumulated over time through research and development, through trial and error in equipment and factory methods, and in the detailed specifications and procedures developed through prolonged experience (Chapter 2).

Of course, some of what is specific to the technological behavior of individual firms may be explained by their competitive strategies as they confront putative rivals and by their response to the characteristics of the national environment in which they operate. Ray Vernon, in his comments on Pavitt's paper at the conference, indeed stressed both of these factors as "powerful determinants of both the nature of innovations that are generated and the channels by which innovations are transferred abroad."

THE POSSIBILITIES AND LIMITS OF TECHNOLOGY TRANSFER

What implications does such an alternative conceptualization hold for the transfer of technology? Insofar as technology is conceived as firm-specific information concerning the characteristics and performance properties of production processes and product designs, and to the extent that it is tacit and cumulative in nature, the transfer of technology is not as easy as the purchase of a capital good or the acquisition of its blueprint. It involves positive and significant resource costs, reflecting the difficult task of replicating knowledge across the boundaries of firms and nations; recipients would normally be obliged to devote substantial resources to assimilate, adapt, and improve upon the original technology. Moreover, the geographic transfer of very productive forms of physical capital may be of little use, unless the appropriate human resources are simultaneously available in situ to provide for the operation, maintenance, repair, and upgrading of the facilities, as well as to interface with and learn from foreign engineers and specialists.

Therefore, to the extent that the normal features of technical knowledge include imperfect understanding, incomplete availability, imperfect imitability, tacitness etc., its successful use tends to be dependent upon firms and countries developing their own technological capabilities. The acquisition of such capabilities is a nontrivial matter. Baranson and Roark, for example, distinguish among technology transfers that impart operational, duplicative, and innovative capabilities and suggest that "none of these capabilities will accrue as a matter of course to passive recipients of technology; each requires an increasing level of technological effort" (Chapter 2).

Much of the effort of recipient countries is geared toward the continuous adaptation of imported technology to local conditions and to the firm's operational characteristics and particular productive constraints. Substantial empirical evidence has in fact accumulated pointing to

> Adaptations [that] have been observed to take place through changes that stretch the capacity of existing plants, break bottlenecks in particular processes, improve the use of by-products, adjust to new input sources, alter the product mix, and introduce a wide variety of incremental improvements in processes and product designs (Westphal, Kim, and Dahlman, Chapter 6).

The circumstances under which technologies are efficiently assimilated tend to be typified by periods of intensive search by the potential buyers, protracted negotiations, aggressive learning tactics, and a strategy focused on adaptation and upgrading of the original design so as to improve its performance characteristics. Such purposive actions on the part of individual firms do not take place easily or costlessly in any environment. As Ozawa stresses in Chapter 7,

> Although Japan's postwar technological miracle would have been impossible had there been no imports of advanced industrial arts from the West, Japanese industry itself exerted a great deal of effort to adapt and assimilate imported technologies. This assimilative effort initially stimulated adaptive R & D, which later turned more original in orientation. There has been a constant rise in R & D expenditures: as a ratio of national income they were a mere 0.8 percent in 1953 but rose to 1.42 percent in 1960, 2.0 percent in 1970, and 2.4 percent in

1980. (The same ratio for the United States was 2.6 percent in 1980 and for West Germany 2.7 percent in 1979.) . . . Similarly, the ratio of royalty payments to R & D expenditures, which had kept rising until 1960, reaching its highest level of 18.5 percent, has thereafter been decreasing: it came down to 5.1 percent in 1980.

In addition, the experiences of the Japanese, and other successful adopters, attest to the importance not only of the industrial but of the social environment as well. In particular, comprehensive educational systems have played major roles in the assimilation of industrial knowledge. Its critical importance in the history of countries that have been successful both in generating and absorbing innovations is one empirical regularity now recognized by most analysts.

> Even before the Meiji Restoration of 1868, which marked the beginning of Japan's modern era, literacy rates (the ability to read public notices, if not literary works) among the key economic decision makers were surprisingly high. . . . One of the policy priorities of the Meiji government for modernization was to introduce a nationwide education system under which all children from 6 through 13 years of age were required to attend school. Human resource development through education, particularly designed to build individuals' capacities and skills needed for modern industrial activities, has ever since been a continuous national commitment (Ozawa, Chapter 7).

Also of great importance, the Japanese have

> been noted for sending personnel to overseas research organs since early in the Meiji period (1868-1914). Around the turn of the century Japanese engineers studied not only at prestigious universities in the United States and Europe, but also at the research laboratories of such firms as Westinghouse and General Electric. Interestingly, this effort to learn by going overseas appears to have intensified, rather than slackening, as Japan has overtaken Western countries in many areas of technology. In 1978 an estimated 40,000 Japanese made trips to the United States to "study and acquire American technology" (Lynn, Chapter 8).

The case of Korea is equally instructive and again attests to the importance of the human resource factor on a country's ability to effectively assimilate and adapt foreign technologies. In a cross-country comparison between the Republic of Korea, Argentina, Brazil, India, and Mexico, it becomes clear that

> What stands out about the educational pattern are the high proportion of postsecondary students abroad, the high secondary enrollment rate, and the high percentage of engineering students among postsecondary students. Even more remarkable is the rapid growth in numbers of scientists and engineers, such that by the late 1970s South Korea apparently had by far the highest percentage of scientists and engineers in the population of the five countries. It likewise appears to have had proportionately more scientists and engineers engaged in R & D and to have spent proportionately more in R & D (Westphal, Kim, and Dahlman, Chapter 6).

In exploring the less well known case of the transfer of textile technology to some of the more industrialized African countries, Mytelka unveils other considerations, which relate to the important phenomenon of "nonlearning," even in the presence of a relatively favorable human and material environment. Increases in operating efficiency over time are hardly an automatic consequence of participation in production. The author argues that learning in production and improvements in productivity appear to be triggered "when persons in key decision-making positions within an enterprise are responsive to their domestic environment, are sufficiently aware of technical possibilities, and are placed in a situation in which there are pressures to reduce production costs" (Chapter 4). When, on the other hand, there are market imperfections and structural rigidities in developing countries and managers of firms find themselves in a less competitive environment, they tend to commit fewer resources to the search for superior technological alternatives. Further, the high search costs in looking outside the firm's "domain of competence" or, alternatively, the increased variance of the results associated with unproven techniques, may also block the adoption of more economical technologies or the modification of the ones in use to suit local conditions better. In the latter case, managers might be trading reduced profits for reduced risks, a problem not uncommon for multinational firms entering an unfamiliar landscape.

MECHANISMS AND MEASUREMENT OF TRANSFERS OF TECHNOLOGY

The transfers of technology have been associated with a number of explicit mechanisms such as direct foreign investment, licensing, consultancy, technical agreements, and turnkey plant and project contracts, as well as trade in capital goods such as machinery, tools, instruments, and transportation equipment. Yet, many of the modes of transfer are hard to detect, let alone measure, given the tacitness and relative openness of technology. Account should be taken, for instance, of the flows of public technological information (scientific and technological journals, patent descriptions) as well as of reverse engineering, the transfer of person- or institution-embodied know-how, visits to production and facilities, and so on (Pavitt, Chapter 1).

The relative importance of each of those mechanisms changes across nations and over time. Baranson notes that in the context of North-South transfer,

> The direct investment package is rapidly giving way to complex joint-venture agreements, production sharing agreements, managing and marketing contracts, service agreements, technology licensing contracts, turnkey and similar contracts, countertrading arrangements, and numerous other forms of nonequity interactions (Chapter 2).

As a recipient of foreign technology, Japan anticipated this trend. Ozawa points out that its industry "secured key technologies mostly under licensing agreements. Direct foreign investments were limited in number and were set up, when they were approved by the government, as joint ventures – with only a handful of exceptions involving whole foreign ownership" (Chapter 7). Another newcomer, the Republic of Korea, presents a not dissimilar picture, with direct foreign investment playing a relatively "minor part in the country's industrialization" whereas in recent years there was an "increased reliance on licensing, . . . explained by the accelerated development of the technologically more advanced industries" (Westphal, Kim, and Dahlman, Chapter 6) such as chemicals, basic metals, and machinery.

Turning the tables, Contractor discusses in detail the circumstances under which firms would prefer licensing as opposed (or in addition) to direct foreign investment. His conclusions are quite relevant to this discussion.

> The ratio of licensing to investment (in a cross-sectional analysis) increases with technical capability in a country, a variable that can indeed be influenced in the long run by government policy, as in Japan. Licensing, however, decreases in relative importance with higher per capita GDP and industrialization, ceteris paribus. . . . As for corporate policy, the absolute level of licensing in a country [is] generally negatively related to the level of direct investment, from which one may conclude that the two are strategy substitutes, rather than complements.
>
> Finally, firms may prefer direct investment under most conditions, but licensing, while remaining minor, may increase its role as the environment is more restrictive toward investment, as the technology is less research intensive, more codified, patented, and transferable, and as the recipient country has greater indigenous technical capabilities (Chapter 9).

THE ROLE OF POLICY

In most countries, the stock of skills, the organizational structures, and the incentive systems necessary to effectively absorb and utilize foreign technologies have often been a product of appropriate policy instruments. The classic success story is that of Japan, and ubiquitous in the literature is the role of the Japanese Ministry of International Trade and Industry (MITI). Lynn notes that this agency has at times used its control over technology imports "to promote the use of advanced technology by Japanese firms, to improve the bargaining position of Japanese firms dealing with foreign suppliers, to facilitate the diffusion of new technology within Japan, and to shape Japanese industrial structure" (Chapter 8). Nevertheless it has not always been successful, and its role has changed markedly over the last three decades toward a more subtle and less openly interventionist one.

In fact, the experiences of many countries reveal a wide variety of outcomes to similar policy experiments of protection of domestic learning, reflecting the complex relation between the transfer of foreign technology and domestic technological development. If it is true, as Lall has pointed out in the case of India, that "a low reliance on imports of technology in the process of industrialization clearly contributes to the buildup of a diverse and deep technological capability" (Chapter 3) it seems to be equally correct to state that protection of both learning and production, if implemented at all costs, brings with it large static and dynamic inefficiencies in the form of outdated technologies, uncompetitive industries, and fragmented markets.

In addition, it must be noted that local efforts in technological creation and upgrading and in the importation of foreign technology more often than not stand in complementary relationship to one another. Even if technological proficiency for certain sectors is achieved with only minor foreign technological involvement, this hardly implies a permanently inverse or substitute relation between the importation of technology and the speed, amount, and quality of local technological learning. On the contrary, the experience of India and other newly industrializing countries suggests that there is some combination of imported and internally generated know-how that may bring the country closest to the desired goals of technological and industrial maturity. Viewing that relationship always in antagonistic terms (the larger our technological imports, the less we learn) may well bring industrial stagnation and technological obsolescence. How to identify that optimal combination is, of course, the difficult question.

Finally, if, with Katz, we define a successful indigenous learning sequence as "one that enables a firm (or an industry) to create and develop a complete set of 'in house' engineering skills – that is, product design skills, process engineering capabilities, and production planning and organization skills – so as to be able to compete on its own both in domestic and in international markets" (Chapter 5), it is normally the case that the completion of this sequence may take as much as 15-20 years.

Under such circumstances, any sensible welfare judgment concerning alternative technological policies and development paths cannot be resolved upon static grounds. What seem to be optimal myopic technological choices might turn out to be third best in face of dynamic gains; conversely, a Pareto-inefficient allocation of resources at any time may be efficient in some growth sense. Thus, a country may not want to follow its *present* comparative advantages in deciding what technologies to adopt, the sequence in which technical capabilities should be acquired, and new activities added or established ones improved.

The case of Japan is again instructive. The technological and industrial choices made by policy makers in the 1950s were apparently "mistaken" by the yardstick of factor proportions and prices. Japan, endowed with an elastic supply of labor and faced with scarcity of capital and natural resources, was actually importing and improving upon capital- and resource-intensive technologies, which were relatively complex and offered substantial economies of scale. A myopic perspective would deem such a course unwise, as it was contrary to Japan's comparative advantage, reflected in its relative factor endowments. Yet, over the longer run, the results of this technological program not only drastically changed Japan's comparative advantage (except in raw materials), but in all probability dominated static or short-run considerations of allocative efficiency.

The success of such policies depends substantially, as was pointed out earlier, on the social and industrial milieu where they are implemented and the nature of the economic agents that are their target. In many cases, an indigenous learning sequence is never successfully completed because of satisficing forms of behavior in a monopolistic environment. In other cases, firms accumulate enough skills and knowledge through technological activities in R & D and production engineering to manufacture goods with correct performance characteristics, but at noncompetitive prices; in spite of approaching the design frontier, plant scales might be insufficient to drive down unit costs to a competitive level.

Still, not enough is known about the relation between technological policy and technological development, given the complex nature of the interactions between policy-induced decisions and the environment where they are implemented, and in light of the extended period during which those actions unfold and results appear. In spite of the recent progress made in understanding those connections, much still remains to be uncovered concerning appropriate policies to foster the international transmission of knowledge and its effective assimilation by recipient countries. The chapters in this volume constitute a significant step in this larger intellectual enterprise.

RESEARCH DIRECTIONS

There remain a number of areas where research should be directed. First, although the importance of an educational and scientific infrastructure and a literate and skilled labor force for the absorption of foreign technologies is now well recognized, the role that should be played by the specific components of an educational system and existing scientific institutions is still far from adequately understood. More precisely, the changing skills and material requirements for an effective transfer of technology in the course of a country's development need to be determined. And how do those requirements change with the level of complexity of the technology and the specific sector to which it is directed?

Second, the nature of the connections between the transfer of foreign technologies and domestic technological development needs to be made more precise. Under what conditions do they hold a complementary relationship and when are they substitutes? In this respect, how does an individual firm's attributes, such as size and nationality, influence that relationship? What is the impact of the character of the market, the number and size distribution of firms, and the degree of competition? And how does the speed with which the technological frontier moves influence the extent to which foreign technologies become essential to indigenous technical development? In addition, are threshold levels of internal development necessary to assimilate foreign technologies, and how may they be characterized? In other words, what are the minimum internal conditions necessary for the absorption of technologies of different degrees of complexity?

Third, what policy instruments should be used to promote an effective transfer of technology and further local development? Insofar as time enters in an essential way, how should policy be formulated to internalize dynamic gains and losses? What should be the role of government on the supply side, in searching for a technology, negotiating its transfer, setting up pilot or demonstration enterprises, and stimulating further upgrading and development? On the other hand, how should procurement policies be structured to optimize the utilization of local resources, foster technical change, and bring about efficient levels of production? Those are not trivial questions. Serious attempts to answer them should provide substantial insights into the continuing problems associated with the international transfer of technology.

Acknowledgments

We are indebted to many scholars whose advice, criticism, and suggestions contributed to this volume. Prominent among these creditors are the members of the Social Science Research Council's Subcommittee on Science and Technology Indicators whose suggestions and criticisms guided the formulation of our initial plans for the conference on which this volume is based. Members of the subcommittee were Jonathan Cole, Gerald Holton, Keith Pavitt, the late Derek de Solla Price, Arnold Thackray, and Harriet Zuckerman, in addition to the senior editor of this volume. We are fortunate to have benefited from the skillful assistance of two council staff associates, Roberta Balstad Miller, who contributed to the initial planning stages, and Robert Pearson, whose quiet competence and unfailing helpfulness were indispensable to the successful completion of the undertaking. We also wish to thank the participants at the conference: Thomas Biersteker, Jennifer Sue Bond, Ronald Findlay, Richard Nelson, Richard Newfarmer, Lois Peters, Rolf Piekarz, Alan Rapoport, Gary Saxonhouse, Gordon Smith, Peter Suttmeier, David Teece, Raymond Vernon, and Dorothy Zinberg. Our greatest debt, of course, is to the authors whose contributions appear in the following chapters of this volume.

The conference was supported by the National Science Foundation under grant no. 77-21686. The opinions, findings, conclusions, and recommendations expressed in this publication are those of the authors and do not necessarily reflect the views of the National Science Foundation.

List of Contributors

Jack Baranson
Illinois Institute of Technology
Chicago, IL 60606

Farok J. Contractor
Graduate School of Management
Rutgers University
92 New Street
Newark, NJ 07102

Carl J. Dahlman
Industrial Strategy and Policy Division
Industry Department
World Bank
1818 H Street, NW
Washington, D.C. 20433

Claudio Frischtak
Industrial Strategy and Policy Division
Industry Department
World Bank
1818 H Street, NW
Washington, D.C. 20433

Jorge M. Katz
Callao 67, 3er. piso/1022
Buenos Aires
Argentina

Linsu Kim
Korea Advanced Institute of Science and Technology
P. O. Box 150
Cheongryang
Seoul
South Korea

Sanjaya Lall
Institute of Economics and Statistics
St. Cross Building
Oxford University
Oxford, OX1 3UL, England

Leonard H. Lynn
Department of Social Science
College of Humanities and Social Science
Carnegie-Mellon University
Schenley Park
Pittsburgh, PA 15213

Lynn K. Mytelka
Department of Political Science
Loeb Building, Room B640
Carleton University
Ottawa, K1S 5B6, Canada

Terutomo Ozawa
Department of Economics
Colorado State University
Fort Collins, CO 80523

Keith Pavitt
Science Policy Research Unit
Sussex University
Falmer
Brighton
Sussex, BN1 9RF, England

Robin Roark
Developing World Industry and Technology, Inc.
800 18th Street, NW, Room 606
Washington, D.C. 20006

Nathan Rosenberg
Department of Economics
Stanford University
Stanford, CA 94305

Larry E. Westphal
Productivity Division
Development Research Department
World Bank
1818 H Street, NW
Washington, D.C. 20433

Contents

List of Tables and Figures

TABLES

FIGURES

Abbreviations

BEA	Bureau of Economic Analysis
BIDI	Banque Ivoirienne de Développement Industriel
BOF	basic oxygen furnace
CAD	computer-aided design
CAM	computer-aided manufacturing
CFA	Communauté Financière Africaine
COMECON	Council for Mutual Economic Aid
DFI	direct foreign investment
EEC	European Economic Community
GDP	gross domestic product
GNP	gross national product
IACs	industrially advanced countries
IBM	International Business Machines
IMF	International Monetary Fund
IRS	Internal Revenue Service
ITC	indigenous technological capability
KILTEX	Kilimanjaro Textiles
KShs	Kenya shillings
MITI	Ministry of International Trade and Industry
MNCs	multinational corporations
MVA	manufactured value added
NICs	newly industrializing countries
OECD	Organization for Economic Cooperation and Development
QC	quality control
R & D	research and development
SIC	standard industrial classification
SONAFI	Société Nationale de Financement
TE	technology exports
TI	technology imports
VOLTEX	Société Voltaïque de Textile
VTR	video tape recorder
WDR	*World Development Report*

INTERNATIONAL TECHNOLOGY TRANSFER

Part I

Technology Transfer in the Context of the Newly Emerging International Division of Labor

1

Technology Transfer among The Industrially Advanced Countries: An Overview

Keith Pavitt

INTRODUCTION

For someone who spends most of his time doing prosaic empirical analysis on the subject of technical change, it was frightening to write a conceptual overview of technology transfer among the industrially advanced countries. Until recently I would have been reluctant to do so. Certainly, empirical studies of technical change have made me uncomfortable with the assumptions that technology is information, that it is embodied in producer goods, or that technology and innovations are generated in high-wage countries and then diffuse with no great difficulty to countries where wages are lower. But the evidence was too scattered, incomplete, and variable to suggest anything different.

In this chapter, I shall begin to do just that, strengthened in my resolve by evidence from a variety of sources: case studies of innovations in the industrially advanced countries and of "technological learning" in the developing countries; more systematic evidence on sectoral patterns of production and use of innovations and on sectoral patterns of comparative technological advantage in different countries; and the contrasting industrial and technological performances of Germany compared to the United Kingdom and of Japan compared to the United States.

I shall restrict myself to the market economies of the Organization for Economic Cooperation and Development (OECD) members since the analysis of centrally planned economies requires special skills and data that I do not possess. I shall also take three things for granted: first, that getting and keeping an innovative lead over competitors is a central feature of capitalist competition, and it is the ability to do this that here explains the advanced countries' comparative advantage in export markets; second, that the assimilation in a country of innovations* from whatever source is an essential ingredient in long-run productivity growth; third, that technology transfer among OECD countries must be seen, since World War II, as part of a process of technological convergence between the United States, on the one hand, and Western Europe and Japan, on the other, that has been highly uneven across countries and across industrial sectors.

Three sets of questions naturally suggest themselves. First, how is technology transferred internationally, and how can we measure the process? Second, what are the proximate explanations for the patterns of international technology transfer that we have observed in the past? Third, what changes and problems can we expect in the future? I shall avoid a fourth set of questions on the normative implications of international technology transfer.

The conceptual tools for answering these questions do not come readily to mind. Assuming that technology is a free good does not help much and is, in any event, contradicted by the growing empirical literature on multinational firms and on the transfer of technology to developing countries. In this chapter, I shall take a different tack, by defining in the next three sections some of the most important characteristics of technology, and the mechanisms for its generation that emerge from empirical studies. I then try to answer the three questions in the light of these characteristics.

It will emerge that the independent technological capabilities of the countries importing technology have probably been the most important factors determining the rate and direction of technology transfer among the OECD countries. The determinants of these national technological capabilities are complex and not well enough understood.

---

*Some brief definitions: *innovation* is a new or better product or process; *technique* is a production system (improved technique is therefore equivalent to innovation); *technology* is knowledge of whatever sort about technique.

TECHNOLOGY IS MAINLY DIFFERENTIATED KNOWLEDGE ABOUT SPECIFIC APPLICATIONS

The specificity of technology emerges from the nature of the inputs to its production and of the resulting outputs. In most advanced countries, at least 60 percent of R & D expenditures are on D, namely expenditures to develop specific products or production processes.[1] In industrial firms themselves, the distribution of the costs of innovation – excluding normal investment in plant and equipment – is roughly as follows: research (R) = 10-20 percent; development (D) = 30-40 percent; production engineering (PE) = 30-40 percent; market launch (M) = 10-20 percent.[2] Depending on the assumptions made, this distribution of expenditures predicts that between 10 and 30 percent of the inputs to industrial technology come from outside industry (mainly universities and government laboratories), and the remainder from within industry itself.

This is indeed what emerges from a number of studies that have tried to identify the institutional sources of knowledge inputs into industrial innovations.[3] These studies also show that a high proportion of technology (between 40 and 60 percent) is generated within the innovating firms themselves, while technology from other firms plays a secondary role. One possible explanation for this pattern is the concern of innovating firms to prevent, by patent secrecy and other means, their technology from being used by their imitators and competitors. This is not completely convincing. A number of studies have also shown that, even when there is no protection of intellectual property, problems of transferring technology remain considerable. The effective use by firms of published knowledge depends on their investment in employing professionally active scientists who are capable of recognizing and assimilating the published knowledge and relating it to the preoccupations of the firm.[4] Even within firms, empirical studies have shown that the transfer of technology involves considerable real costs to the user.[5]

TECHNOLOGY IS OFTEN UNCODIFIED, AND LARGELY CUMULATIVE WITHIN FIRMS

Behind the large contribution to innovation of technology generated within the innovating firm, and the high costs of absorbing technology from outside, lies a highly differentiated range of techniques and related technological knowledge. These cannot easily be derived from or reduced to first scientific principles; they depend therefore on a range of acquired skills, practices and subtheories (or rules of thumb). The acquisition of technology is always involved when a firm moves from one vintage of production technique to another, or from one product group to another. Such acquisition involves not only written information (for example, patents, blueprints, operating instructions), but also person-embodied skills and know-how and the adaptation of techniques to local operating conditions and markets.

We are therefore a long way from the concept of technology as information that is generally applicable and easy to reproduce and re-use[6] and where firms can produce and use innovations mainly by dipping freely into a general stock or pool of technological knowledge. Instead we have firms producing things in ways that are differentiated technically from things in other firms, and making innovations largely on the basis of in-house technology, but with some contribution from other firms, and from public knowledge. Under such circumstances, the search process of industrial firms to improve their technology is *not* likely to be one where, assuming a general pool of technological knowledge, they look over the whole range of that knowledge before making their technical choices. Instead, given its highly differentiated nature, firms will seek to improve and to diversify their technology by searching in zones that enable them to use and to build upon their existing technological base. In other words, technological and technical changes in the firm are cumulative processes. What the firm can hope to do technologically in the future is heavily constrained by what it has been capable of doing in the past.

There is perhaps a useful metaphor between technology and sports practice. Neither can be performed satisfactorily simply through an understanding of physical laws, or through reading books. Improvement comes mainly through practice, especially against those who are somewhat better, but rarely against those who are overwhelmingly better. The learning or diffusion of sport seldom involves complete replication; there is always some adaptation in style or method to meet specific conditions. In addition, there are several sports (technologies), each with differing sizes, shapes, and lengths of ball, racket, pitch, track, or team. Proficiency in one sport does not guarantee proficiency in another, but these will be more or less closely related. Thus, being good at badminton may help in being good at squash, but is not likely to help in rugby, where being good at American football is likely to be a better guide.

SOURCES AND PATTERNS OF TECHNOLOGICAL CHANGE VARY AMONG SECTORS

For similar reasons we see that, at any one time, firms have different patterns of "technological diversification" across product groups: in other words, they have differing ranges of product groups in which they produce technology and innovations. These patterns reflect cumulative processes where, from differing core activities, firms have followed over time differing technological paths (or trajectories) into related technological zones. These differing patterns are, I have argued,[7] at the heart of observed differences among sectors in the contribution to innovation by firms of different sizes and of observed sectoral patterns of production and use of innovations. They also have implications for channels of international technology transfer and for the mechanisms that induce the accumulation of technological skills in firms and countries.

Table 1.1 attempts to classify the variety of patterns of innovation into three categories of firm and identifies the typical core activity of such firms. It describes the varying patterns of innovation in terms of product/process innovation, size of innovating firm, and degrees and patterns of technological diversification across product groups. These differences can be explained, as proposed by Nelson,[8] by three sets of characteristics related to each category of firm: sources of technology, type of customer, and means available to the firm of appropriating technological advantage.

Thus, in supplier-dominated firms, there is relatively little technological capability in the typically small firms beyond that required to operate and occasionally improve upon machines provided by suppliers. Customers are usually price-sensitive users, so that firms focus on process innovation. Firm-specific appropriation is generally through advertising and marketing, trademarks, and aesthetic design, rather than through technology. In the sectors made up of such firms, technological advance is likely to be a function of wage levels and of autonomous technical improvements in the production

TABLE 1.1
Patterns of Innovation and Type of Firm

Patterns of Innovation	*Category of Firm*			
	Supplier-dominated	*Scale-intensive or Production-intensive*	*Specialized suppliers*	*Science-based*
Typical core sectors	Agriculture; housing; private services; traditional manufacture	Bulk materials (steel, glass); assembly (consumer durables & auto	Machinery; instruments	Electronics/electrical; chemicals
Sources of technology	Suppliers; research & extension services; big users	PE; suppliers; R & D	Design & development; users	R & D; public science; PE
Type of user	Price sensitive	Price sensitive	Performance sensitive	Mixed
Means of appropriation	Nontechnical (i.e., trademarks, marketing, advertising, aesthetic design)	Process secrecy & know-how; technical lags; patents; dynamic learning economies	Design know-how; knowledge of users; patents	R & D know-how; patents; process secrecy & know-how; dynamic learning economies
Balance between product & process innovation	Process	Process	Product	Mixed
Size of innovating firms	Small	Large	Small	Large
Technological diversification	Low, upstream	High, upstream	Low, concentric	Low, upstream; high, concentric

Source: Derived from K. Pavitt, "Patterns of Technical Change: Towards a Taxonomy and a Theory."

machinery supplying the sector. International technology transfer takes place mainly through trade in such machinery.

Over time, firms in some sectors evolve into the second, production-intensive class. To some extent, this reflects a mechanism, already described by Adam Smith, linking the extent of the market, the division of labor, the simplification of tasks, and the substitution of machines for labor. It also reflects continuous technical improvements in capital goods and the exploitation of the economies of scale inherent in continuous process production.

In such firms, the exploitation of economies of scale is a continual preoccupation with selling to what are largely price-sensitive users. Improvements in process technology are therefore of central importance. But these cannot be left simply to the machinery suppliers, as is the case in supplier-dominated firms. Production systems are large, complex and interdependent. Their reliability and economic operation have become the subject of continuing professional attention from process or production engineering departments, which in turn have learned how to improve and change the production system and to design and build related production equipment. These departments live in symbiosis with generally small and specialized machinery suppliers, who benefit from the facilities of other large customers in developing, testing, and operating their equipment, and who can offer them in return the advantages of equipment designed and used over a wide range of firms and sectors.

In the production-intensive sectors, trade in production machinery therefore reflects only part of international technology transfer. Skills and know-how in the design, construction, and operation of large-scale production systems are of equal or greater importance, and these are likely to be reflected in payments for know-how found in licensing agreements or in transactions within multinational firms.

However, neither licensing agreements nor foreign direct investment appear to be the major channels of international transfer of machine-building technology. Product design and reliability in response to users' needs are the critical factors in competitive success, and these can rarely be protected efficiently through patents or secrecy. Reverse engineering* and linkages with production

*Reverse engineering is essentially imitation and adaptation without any formal agreement with the innovator. It involves stripping down innovative products and finding out how they work.

engineering departments in scale-intensive sectors are likely to be of central importance in international technology transfer; the latter will often involve "local content" agreements with foreign-owned firms producing in the country.

The rate and dircction of technical change in scale-intensive firms in different countries will depend not just on the level of wages but also on other significant input costs (for example, raw materials and energy in continuous process industries); on the prospective size of the market and the consequent possibilities of exploiting economies of scale; on the quantity and quality of resources devoted to improvements in technology and technique in the production of process engineering departments in scale-intensive firms; and in the design and development departments of the specialized firms supplying production equipment.

Similarly, in the supply of specialized machinery, high or increasing wages are unlikely to be the sole inducement to technical change. Linkages with production engineering departments in scale-intensive firms are also likely to be important, as are linkages resulting from favorable local endowments of natural resources that require specialized capital goods for their exploitation (for example, coal, oil, metal ore, food, textiles, jute).

In science-based firms, the paths of technological development are dictated by powerful techniques that have emerged from the underlying sciences: chemical (and now biological) synthesis in chemical firms; electromagnetism, radio waves, and the transistor effect in electrical/electronic firms. Technology draws heavily on industrial R & D activities and (more than in other sectors) on public scientific and technical knowledge produced in the universities and elsewhere. Potential applications are pervasive: from plastics and televisions, where process innovation is important; to pharmaceuticals and process control equipment, where product innovation and reliability are important. Firms mastering these science-based technologies can exploit a rich vein of product and process opportunities that are based on them.

More than in other industries, technology transfer depends on the user's understanding of the underlying scientific and technical principles. Industrial R & D will therefore be an essential component of the capacity to assimilate foreign technology. Given that much outside knowledge flows into industry through people, important channels of transfer are local basic research and the training of scientists and engineers in foreign academic institutions. Financial payments for technology will be of major importance in two areas: first, for production know-how when exploiting the science-based technology involves large-scale production (principally in bulk synthetic materials and in consumer electronics); second, for patent licenses in pharmaceutical and other fine chemicals, where the technical and know-how barriers to imitation are negligible.

The national location of innovative activity will depend not just on the wage rate, but also on the capacity to exploit the technological opportunities emerging from the underlying sciences: R & D activities in firms and related basic research activities in universities will therefore be particularly important.

EVALUATING INTERNATIONAL TECHNOLOGY TRANSFER

To sum up, technology is mainly specific knowledge about highly differentiated products and production processes that accumulates step-by-step in firms. The successful assimilation of technology from outside sources depends on an in-house capacity not just in R & D but also in production engineering. Assimilation invariably involves adaptation, so that the diffusion of an innovation cannot be neatly separated from innovation itself. Given these characteristics, we can try to answer our first question, How is technology transferred internationally, and how can we measure the process?

The sources, nature, and mechanisms of international transfer of technology vary considerably from sector to sector. In sectors where firms are in general supplier-dominated, technology comes mainly already embodied in production machines. In production-intensive firms, the key technology relates to constructing and operating large-scale plant and is transferred internationally mainly through know-how agreements. In sectors supplying production equipment, however, technology is transferred internationally mainly through reverse engineering and through local linkages with the production engineering departments in production-intensive user firms. In science-based firms, the key technology emerges mainly from industrial R & D and in some cases from academic research. Channels of international technology transfer are various, including patenting and know-how licensing, reverse engineering, and person-embodied transfers through academic institutions. Given these mechanisms of international technology transfer, its measurement is up against six difficulties.

First, the international flows of public technological information (scientific and technological journals, patent descriptions) and of firm-specific information (drawings and blueprints, operating manuals), are only part of the process of international technology transfer. They do not capture reverse engineering, the transfer of person- or institution-embodied know-how, flows through academic institutions, or technology embodied in machines.

Second, monetary flows resulting from inter- or intra-firm licensing agreements can, in principle, capture flows of know-how. Patent licenses can be distinguished from know-how licenses. Within the latter it is useful to distinguish between sums involved in transferring otherwise secret information and those involved in transferring person-embodied know-how.

Third, for certain important channels of international transfer, we have poor information. In trade in production machinery, it is difficult to separate out the value of embodied technology. Data on the training of foreign students have been collected, but not principally for the purposes of measuring international technology transfer.

Fourth, on imitation through reverse engineering we have no systematic information at all. As has been seen, the assimilation of outside technology always involves considerable costs to those doing the assimilating, and requires activities that resemble in many ways those necessary for the creation of technology. A good proxy measure for the capacity for imitation is therefore the level of industrial activity in R, D, and production engineering (PE). Although there are good data on R & D activities in the OECD countries, there is nothing systematic on PE activities.

Fifth, we must be wary of partial measures of the value and relative importance of technology imports. For example, I and others have in the past understood technology license payments to be proxies for technology imports and industrial R & D expenditures to be proxies for the production of indigenous technology. This view is mistaken. There are many other channels for the import of foreign technology than licensing, and industrial R & D activities not only create indigenous technology but also assimilate and adopt technology of foreign origin.

Finally, even if better data are collected on each of the channels of international technology transfer, there remains the problem of putting them together on a common measure of volume and value.

These very real difficulties should not be allowed to discourage data collection and interpretation. Progress can be and has been made through information collected from other sources.[9]

INTERNATIONAL TECHNOLOGY TRANSFER IN THE PAST

The second question that we have set out to answer is, What are the proximate explanations for the patterns of international technology transfer that we have observed in the past? One set of explanations concentrates on the international diffusion of best practice U.S. technology in the period following World War II.[10] Best practice technology in the United States, it is argued, was stimulated by and allowed high wages. Other countries were able to catch up with the United States given two characteristics of the OECD's economic system: first, growing economic interdependence that opened and widened the channels of international technology transfer through trade in producer goods, licensing agreements and direct foreign investment, and that allowed a fuller exploitation of economies of scale in production; second, high rates of investment in the catching-up countries that encouraged a rapid rate of diffusion of best practice technique in the capital stock.

The discussion earlier in this chapter does not contradict such a view but complicates it by adding a number of dimensions. They concern the level of technological capability in the countries importing the technology and the long periods over which these accumulate. They also concern the diversity of mechanisms through which firms and countries can accumulate a technological advantage. These help to explain why the catching-up process has been unevenly distributed among the industrially advanced countries, and why in some sectors other countries have overtaken the United States in technology, technique, and productivity.

One characteristic of the countries catching up with U.S. technological levels in the postwar period is their relatively high level of accumulated indigenous technology, reflected in industry-financed R & D activities and in industry-related skills. I argued above that levels of indigenous technology are a good proxy measure of the capacity to assimilate technology from outside sources. So have various students of Japan's very successful postwar record of technology assimilation.[11] The Japanese experience suggests not only that assimilation and diffusion is rarely passive, but also that it need not stop once the world best practice technology is reached. Assimilative R, D, and PE activities can evolve toward innovative R, D, and PE. This has been the Japanese experience in sectors such as shipbuilding, iron and steel, and consumer electronics.[12] In other respects, however, Japan is not typical of the industrially advanced countries that have closed the technological and productivity gap with the United States over the past 35 years. If one leaves the centrally planned countries out of consideration, Japan is the only country over the past 60 years that has joined the relatively small and select group where significant numbers of world frontier innovations are developed and produced.[13] Japan's eventual emergence as a powerful innovating country can be seen as the continuation of a vigorous policy of technological accumulation begun more than a century ago. The other major OECD countries had strong capacities for world frontier innovations *before* World War II. In particular, the technological dynamism of Germany, compared to the United Kingdom, became visible by the beginning of this century, was noticed at the time by Marshall,[14] and has been documented since by Allen,[15] Hobsbawm,[16] and Landes.[17] The conventional wisdom among many policy makers today remains that the United Kingdom is strong in basic science, invention, and radical innovation but weak in commercial and production follow-through.[18] In my view, such an assessment reflects the special circumstances of the period from 1945 to 1960, when the Federal Republic of Germany and Japan, among others, were preoccupied with physical reconstruction. If one takes a longer view back to the beginning of this century, the German contribution to basic science, invention and radical innovation has been outstanding, not only in chemicals, machine building, and steel, but also in the concepts and technologies underlying modern aviation.[19] One is tempted even to suggest that Germany's contribution to industrial technology in the twentieth century has been greater than that of the United States. Even if the weight of evidence turns out against such a hypothesis, the very considerable German contribution (and that of some of its close neighbors, in particular, the Netherlands, Sweden, and Switzerland) to industrial technology has to be explained.

Davidson[20] has suggested that other factor scarcities, in addition to labor scarcities, can stimulate innovative activities in a country. In particular, scarcities of land, as well as lack of assured supply of raw materials, have been powerful stimuli to certain types of innovation in Europe and Japan: for example, in synthetic materials, large-scale process technology, and small automobiles. In this chapter I proposed a number of other factors around which technological capabilities accumulate in a country, beyond scarce labor and natural resources:

1. private and public investment activities that not only diffuse best practice production techniques but also stimulate innovation in upstream equipment suppliers: for example, *abundant* natural resources (mining machinery in Sweden and the United Kingdom), railroads (France), defense equipment (France, the United Kingdom, the United States).
2. accumulating technological skills in R, D, and PE that enable firms to develop and maintain a firm-specific competitive advantage in a related but evolving range of product groups (chemicals and machinery in West Germany, electronics in the Netherlands, machinery in Sweden, fine chemicals in Switzerland, electronics in Japan).

I therefore conclude that international patterns of technological convergence toward the world best practice technology since World War II reflect long-standing international patterns of technological accumulation in the assimilating countries, and that these cannot be explained simply as a function of wages.

INTERNATIONAL TECHNOLOGY TRANSFER IN THE FUTURE

This discussion of the past patterns of international technology transfer enables us to approach the third question, What changes and problems can we expect in future? As someone who was involved in the analyses and discussions in the mid-1960s about the technological gap between the United States and Western Europe, I am acutely aware that perceptions about trends in the technological capabilities of countries can be wrong.[21] What follows can therefore be no more than speculation.

In one scenario there will be no great problems. Western Europe and Japan have become increasingly Americanized, with large markets and high wages, while the United States has become increasingly Europeanized, with high-cost energy and other resources. To the extent that the volume and direction of R, D, and PE in industrial firms are sensitive to the immediate structure of demand and costs, we can expect greater homogeneity and equality in the distribution of innovative activities among countries and in patterns of international technology transfer. At the same time, some sectoral differences in national technological advantage will remain, given variations in natural resource endowments and in the intensity and direction of programs of public investment.

But what if the volume and direction of R, D, and PE in countries are *not* sensitive to the structure of demand and costs? I have argued above that the technological paths and trajectories of firms result from a cumulative process that, although obviously related to markets in some general sense (for example, achieving economies of scale in production; ensuring that product performance is adapted to user needs), also has its own *technological* logic that both constrains and focuses the search behavior of firms. Eventually, the rate and direction of R, D, and PE might be made dependent on a range of other national variables influencing firm behavior: for example, profitability, the cost of money, the performance criteria of financial institutions, the training and behavior of managers, the training and behavior of workers, the size of national and government markets. For the moment, however, any stated relationship between these variables and national differences in levels and trends in R, D, and PE comes from anecdote or prejudice rather than from systematic evidence.

Once R, D, and PE in a country are made an independent variable, future scenarios become much more slippery and potentially difficult. Countries can catch up fast and overtake, as West Germany and Japan have done, or they can fall behind the frontier as the United Kingdom has done. I shall therefore explore a scenario that I consider plausible over the next ten years:

1. The United States maintains its lead in technologies closely related to military activities. This poses some problems in international technology transfer West/West or West/East in such areas as aircraft, space, nuclear materials, and electronics, but these are similar to problems that have existed over the past 25 years.

2. No new countries join the select club at the world technological frontier. The newly industrializing countries continue to assimilate foreign technology through machinery imports, production know-how licenses, and local contract agreements in supplier-dominated and scale-intensive sectors. Their indigenous technological capacity in PE and related production machinery will develop, but their industrial R & D will remain negligible.

3. Japan, West Germany, and some of its neighbors maintain their technological dynamism and increase their innovative capacity in general and in their particular sectors of comparative advantage, namely, chemicals (West Germany and Switzerland), automobiles (West Germany, Sweden, and Japan), electronics (Japan), and production machinery (West Germany, Japan, Sweden, and Switzerland).

4. Assume that this technological dynamism has no great ramifications in the chemical industry, given the existence of technologically strong companies in other countries (principally the United Kingdom and the United States). Assume, further, that it does have strong ramifications in automobiles, consumer electronics, and production machinery, given that firms in the rest of the OECD countries are technologically and competitively relatively weak.

A key question then becomes whether or not this technological gap will be self-sustaining or even cumulative, or whether it will be closed, just as the technological gap between the United States and other OECD countries has been closed in the past. At first sight, it seems likely that it will be more difficult to close, for three interrelated reasons: first, the way in which West German and Japanese firms exploit their technological advance on world markets; second, the nature of the technologies in which they have a strong advantage; third, the nature of the skills related to the technologies.

One characteristic of the period of intensive technology transfer between the United States and Western Europe was that U.S. firms transferred technology mainly through direct investment and licensing, rather than through direct exports. This is in contrast to the

Federal Republic of Germany and Japan, where exports have been predominant.[22] For all but our supplier-dominated firms (see Table 1.1), licensing and foreign direct investment transfer more technology than trade in machinery. A number of reasons have been put forward to explain this behavior of West German and Japanese firms. I wish to suggest an explanation where, in terms of Dunning's eclectic theory of foreign investment,[23] there is a nice complementarity between firm-specific technological skills and country-specific skills in the Federal Republic and Japan.

In the sectors of West German and Japanese technological strength (consumer electronics, automobiles, and production machinery), firm-specific advantages are strongly developed activities in design and in production engineering. These dovetail nicely with a set of country-specific advantages: a well-trained and disciplined labor force; a plentiful supply of well-trained engineers and technicians; and local component and equipment suppliers that are technically advanced and reliable on deliveries. Fitting these two sets of advantages together enables products of high quality and the full exploitation of potential learning economies. Given that systems of industrial training, trade union structures and the web of suppliers are difficult to include in any package of international technology transfer, important dimensions of a capacity to assimilate foreign technology will be lacking if there is no national policy in the receiving country to change or improve things.

However, there are some countervailing tendencies. A cumulative competitive advantage based on exports will feed back through relative exchange rates to make direct foreign investment more attractive. So will threats of protection reflecting concern about unemployment. At the same time, unemployment will make trade unions in technology-importing countries more amenable to change to the ways of West German and Japanese management, while familiarity and exposure appear to increase mutual respect.[24] West German and Japanese firms might also make the export of their production know-how a special line of business. Finally, technical change in production methods, and particularly in the development in and diffusion of flexible manufacturing systems, may increase the potential attractiveness of foreign direct investment by decreasing the scale and increasing the flexibility of economically viable production systems.[25]

CONCLUSION

This chapter has not succeeded in presenting a comprehensive, coherent, and rigorous conceptual framework that measures and explains patterns of international technology transfer among the OECD countries. It has, however, proposed two axioms that have important implications for the ways in which patterns of international technology transfer are measured and interpreted. First, technology is not information that is generally applicable and easily transferred and used; it is specific to applications and accumulates in firms along paths that reflect a technological logic. Second, the vigor with which firms and countries move along technological paths depends not just on factor scarcities, but also on levels and patterns of public and private investment and on the autonomous decisions of firms to put more or fewer resources into research, development, and production engineering activities.

Under such circumstances, it cannot be assumed that international technology transfer among the industrially advanced countries will necessarily lead to their convergence at the best practice world technology frontier. Such convergence in the postwar period had two special characteristics: first, Western Europe and Japan had already accumulated considerable technological knowledge and skills that enabled them to assimilate U.S. technology effectively; second, a high proportion of U.S. technology was exported through direct foreign investment rather than through exports. Neither of these conditions holds to the same extent in relation to any West German or Japanese technological lead in automobiles, consumer electronics, and production machinery, where the leads may become cumulative.

At the center of these potential problems lie differences among countries in their rate of accumulating technological skills and knowledge. Considerable progress has been made over the past 20 years in measuring countries' technological activities and international transfers of technology. Some progress has also been made in explaining different countries' sectoral patterns of technological advantage. But we know hardly anything systematic about the factors behind the different overall volume and trends in national innovative activities.

Our concepts of technology, its generation, and its diffusion also suggest some of the properties that a formal model should have, if it is to approximate usefully to reality. Above all, it should *not* assume that technology has the properties of information, nor should it take technology as a public good as the norm. At its core should be the process of the creation, international diffusion, and adoption and adaptation of firm-specific innovations. The rate and direction of creation, the channels of international diffusion, and the rate of adoption and adaptation should all be made dependent on a range of variables, many of which have been identified in this chapter.

NOTES

1. National Science Board, *Science Indicators: 1980* (Washington, D.C., 1981).

2. J. Kamin, I. Bijaoui and R. Horesh, "Some Determinants of Cost Distribution in the Process of Technological Innovation," *Research Policy* 11 (1982), pp. 83-94.

3. J. Langrish, M. Gibbons, W. Evans and F. Jevons, *Wealth from Knowledge* (London: Macmillan, 1972); M. Gibbons and R. Johnstone, "The Roles of Science in Technological Innovation," *Research Policy* 3 (1974), pp. 220-242; J. Townsend et al., *Innovations in Britain since 1945*, Occasional Paper no. 16, Science Policy Research Unit (Brighton: University of Sussex, 1981).

4. Gibbons and Johnstone, "The Roles of Science in Technological Innovation."

5. D. Teece, *The Multinational Corporation and the Resource Cost of International Technology Transfer* (Cambridge, Mass.: Ballinger, 1977); D. de Melto, K. McMullen and R. Wills, *Innovation and Technological Change in Five Canadian Industries*, Discussion Paper no. 176, Economic Council of Canada (Ottawa, 1980).

6. K. Arrow, "Economic Welfare and the Allocation of Resources for Invention," *The Rate and Direction of Inventive Activity* edited by R. Nelson (Princeton, N.J.: Princeton University Press, 1962), pp. 609-625.

7. K. Pavitt, "Sectoral Patterns of Technical Change: Towards a Taxonomy and a Theory," *Research Policy* 13 (1984), pp. 343-373.

8. R. Nelson, *The Moon and the Ghetto* (New York: Norton, 1977).

9. De Melto et al., *Innovation and Technological Change*, p. 11; Townsend et al., *Innovations in Britain*; K. Pavitt, "Some Characteristics of Innovative Activities in British Industry," *Omega* 2 (1983), pp. 223-243; W. Davidson, "Patterns of Factor Saving Innovation in the Industrialized World," *European Economic Review* 8 (1976), pp. 207-217; R. Vernon and W. Davidson, *Foreign Production of Technology-Intensive Products by U.S.-Based Multinational Enterprises* (Washington, D.C.: National Science Foundation, 1979).

10. Vernon, "International Investment and International Trade in the Product-Cycle," *Quarterly Journal of Economics* 80 (1966), pp. 190-207; B. Williams, *Technology, Investment and Growth* (London: Chapman Hall,

1967); J. Cornwall, *Modern Capitalism, Its Growth and Transformation* (London: Martin Robertson, 1977); Organization for Economic Cooperation and Development, *Technical Change and Economic Policy* (Paris: OECD, 1980).

11. K. Oshima, "Research and Development and Economic Growth," *Science and Technology in Economic Growth* edited by B. Williams (London: Macmillan, 1973); M. Peck and A. Goto, "Technology and Economic Growth: The Case of Japan," *Research Policy* 10(3) (July 1981), pp. 222-243.

12. M. Peck and R. Wilson, "Innovation, Imitation and Comparative Advantage: The Performance of Japanese Colour Television Set Producers in the U.S. Market," *Emerging Technologies: Consequences for Economic Growth, Structural Change and Employment* edited by H. Giersch (Tübingen: J.C.B. Mohr, 1982), pp. 195-214.

13. K. Pavitt and L. Soete, "International Differences in Economic Growth and the International Location of Innovation," *Emerging Technologies* edited by H. Giersch, pp. 105-133.

14. A. Marshall, *Industry and Trade* (London: Macmillan, 1919).

15. G. Allen, *The British Disease: A Short Essay on the Nation's Lagging Wealth* (London: Institute for Economic Affairs, 1976).

16. E. Hobsbawn, *Industry and Empire* (London: Penguin, 1968).

17. D. Landes, *The Unbound Prometheus* (Cambridge: Cambridge University Press, 1969).

18. S. Feinman and W. Fuentevilla, *Indicators of International Trends in Technological Innovation* (Washington, D.C.: National Science Foundation, 1978); Pavitt, "Technology in British Industry: A Suitable Case for Improvement," *Industrial Policy and Innovation* edited by C. Carter (London: Heinemann, 1981), pp. 88-115.

19. R. Miller and D. Sawers, *The Technical Development of Modern Aviation* (London: Routledge & Kegan Paul, 1968).

20. W. Davidson, "Patterns of Factor Saving Innovation."

21. Organization for Economic Cooperation and Development, *Gaps in Technology: General Report* (Paris, OECD, 1968).

22. S. Hymer and R. Rowthron, "Multinational Corporations and International Oligopoly: The Non-American Challenge," *The International Corporation* edited by C. Kindleberger (Cambridge, Mass.: MIT Press, 1970), pp. 57-91.

23. J. Dunning, "Towards an Eclectic Theory of International Business: Some Empirical Tests," *Journal of International Business Studies* 11 (1980), pp. 9-31.

24. W. Reitsberger, "Japanese Manufacturing in Europe: Myths and Realities," paper presented at the eighth annual conference of the European International Business Association, INSEAD, Fontainebleau, France (1982).

25. A. Altshuler, M. Anderson, D. Jones, D. Roos and J. Womack, *The Future of the Automobile* (Cambridge, Mass.: MIT Press, 1984).

REFERENCES

G. Allen. *The British Disease: A Short Essay on the Nation's Lagging Wealth.* London: Institute for Economic Affairs, 1976.

A. Altshuler, M. Anderson, D. Jones, D. Roos and J. Womack. *The Future of the Automobile*. Cambridge, Mass.: MIT Press, 1984.

K. Arrow. "Economic Welfare and the Allocation of Resources for Invention." *The Rate and Direction of Inventive Activity* edited by R. Nelson. Princeton, N.J.: Princeton University Press, 1962, pp. 609-625.

J. Cornwall. *Modern Capitalism, Its Growth and Transformation*. London: Martin Robertson, 1977.

W. Davidson. "Patterns of Factor Saving Innovation in the Industrialised World." *European Economic Review* 8 (1976): pp. 207-217.

D. de Melto, K. McMullen and R. Wills. *Innovation and Technological Change in Five Canadian Industries*. Discussion Paper no. 176. Ottawa: Economic Council of Canada, 1980.

J. Dunning. "Towards an Eclectic Theory of International Business: Some Empirical Tests." *Journal of International Business Studies* 11 (1980): pp. 9-31.

S. Feinman and W. Fuentevilla. *Indicators of International Trends in Technological Innovation*. Washington, D.C.: National Science Foundation, 1978.

M. Gibbons and R. Johnstone. "The Roles of Science in Technological Innovation." *Research Policy* 3 (1974): pp. 220-242.

E. Hobsbawn. *Industry and Empire*. London: Penguin, 1968.

S. Hymer and R. Rowthorn. "Multinational Corporations and International Oligopoly: The Non-American Challenge." *The International Corporation* edited by C. Kindleberger. Cambridge, Mass.: MIT Press, 1970, pp. 57-91.

J. Kamin, I. Bijaoui and R. Horesh. "Some Determinants of Cost Distributions in the Process of Technological Innovation." *Research Policy* 11 (1982): pp. 83-94.

D. Landes. *The Unbound Prometheus*. Cambridge: Cambridge University Press, 1969.

J. Langrish, M. Gibbons, W. Evans and F. Jevons. *Wealth from Knowledge*. London: Macmillan, 1972.

A. Marshall. *Industry and Trade*. London: Macmillan, 1919.

R. Miller and D. Sawers. *The Technical Development of Modern Aviation*. London: Routledge & Kegan Paul, 1968.

National Science Board. *Science Indicators: 1980*. Washington, D.C., 1981.

R. Nelson. *The Moon and the Ghetto*. New York: Norton, 1977.

Organization for Economic Cooperation and Development. *Gaps in Technology: General Report*. Paris: OECD, 1968.

——. *Technical Change and Economic Policy*. Paris: OECD, 1980.

K. Oshima. "Research and Development and Economic Growth." *Science and Technology in Economic Growth* edited by B. Williams. London: Macmillan, 1973, pp. 310-323.

K. Pavitt. "International Technology and the US Economy: Is There a Problem?" *The Effects of International Technology Transfers on US Economy* edited by R. Piekcarz. Washington, D.C.: National Science Foundation, 1974, pp. 61-76.

——. "Some Characteristics of Innovative Activities in British Industry." *Omega* 11 (1983): pp. 113-130.

——. "Technology in British Industry: A Suitable Case for Improvement." *Industrial Policy and Innovation* edited by C. Carter. London: Heinemann, 1981, pp. 88-145.

K. Pavitt and L. Soete. "International Differences in Economic Growth and the International Location of Innovation." *Emerging Technologies: Consequences for Economic Growth, Structural Change and Employment* edited by H. Giersch. Tübingen: J. C. B. Mohr, 1982, pp. 105-133.

K. Pavitt. "Sectoral Patterns of Technical Change: Towards a Taxonomy and a Theory." *Research Policy* 13 (1984): pp. 343-373.

M. Peck and A. Goto. "Technology and Economic Growth: The Case of Japan." *Research Policy* 10(3) (July 1981): pp. 222-243.

M. Peck and R. Wilson. "Innovation, Imitation and Comparative Advantage: The Performance of Japanese Colour Television Set Producers in the U.S. Market." *Emerging Technologies: Consequences for Economic Growth, Structural Change and Employment* edited by H. Giersch. Tübingen: J.C.B. Mohr, 1982, pp. 195-214.

W. Reitsberger. "Japanese Manufacturing in Europe: Myths and Realities." Paper presented at the eighth annual conference of the European International Business Association, INSEAD, Fountainebleau, France, 1982.

N. Rosenberg. *Perspectives on Technology*. Cambridge: Cambridge University Press, 1976.

L. Soete. "A General Test of Technological Gap Trade Theory." *Review of World Economics* 117 (1981): pp. 638-666.

D. Teece. *The Multinational Corporation and the Resource Cost of International Technology Transfer*. Cambridge, Mass.: Ballinger, 1977.

J. Townsend, F. Henwood, G. Thomas, K. Pavitt and S. Wyatt. *Innovations in Britain since 1945*. Occasional Paper no. 16. Brighton: University of Sussex, Science Policy Research Unit, 1981.

R. Vernon. "International Investment and International Trade in the Product Cycle." *Quarterly Journal of Economics* 80 (1966): pp. 190-207.

R. Vernon and W. Davidson. *Foreign Production of Technology-Intensive Products by U.S.-Based Multinational Enterprises*. Washington, D.C.: National Science Foundation, 1979.

B. Williams. *Technology, Investment and Growth*. London: Chapman Hall, 1967.

2

Trends in North-South Transfer of High Technology

Jack Baranson
and Robin Roark

INTRODUCTION

Transfers of high technology from the industrially advanced countries (IACs) in the North to certain developing countries in the South have continued and increased in recent years in spite of the widespread recognition that such technology may not be appropriate to the needs and resources of the recipient country in many cases. This chapter examines some patterns and trends in this transfer. The first section provides a definition of *high technology* and a short discussion of the problem of measuring its transfer. This effort is problematic at best but serves to limit the scope of discussion somewhat. The second section examines trends that describe the shift toward high technology by newly industrializing countries (NICs) and developing countries. The third section examines patterns of response in the IACs to demand for high technology on the part of developing countries. The final section summarizes problems and issues implicit in North-South high technology transfer.

HIGH TECHNOLOGY DEFINED AND MEASURED

Technology and *technology transfer* have been defined in many ways. In the simplest view, technology is a pool of freely available information used to produce goods of value. Yet the package of product designs, input processing and production techniques, managerial systems, marketing rights, and technical knowledge commonly included in a technology transfer is not a public good. Significant costs are borne by both technology suppliers and recipient organizations in such transfers.

The technology package also consists of detailed progress sheets, materials specifications, processing and testing equipment designations, and quality control procedures. Together they are a measure of the quantum and complexity of a technology transfer. The tens of thousands of elements required for a single industrial product, such as a high-speed diesel engine, are meticulously accumulated over time through research and development, through trial and error in equipment and factory methods, and in the detailed specifications and procedures developed through prolonged experience. The quantum and complexity of technology in the chemical processing and other high-technology fields, such as computers and jet aircraft, are even greater.

The dynamics of technological change preclude a final delineation of specific goods or industries that can be considered as high technology. The search for an appropriate U.S. policy for international technology transfers has given rise to several studies that devised methods of measuring the technology intensity of trade. Two widely used proxies for the technology intensity of firms or industries are: 1) the number of scientists and engineers engaged in research and development as a percentage of the firm or industry workforce, and 2) expenditures on R & D as a percentage of value added or total sales.

Since the mid-1970s, techniques for measuring technology intensity have been widely debated, but three have come to the forefront: one is an industry-based definition and two are product based. In each, products (or industries) are disaggregated (to the extent possible) according to standard industrial classification (SIC) numbers and ranked according to technology intensity as measured by various proxies.* Usually, an arbitrary cut-off point is chosen to separate

---

*Boretsky developed an industry-based definition of technology intensity using R & D expenditures and employment of scientists, engineers, and technicians as proxy measurements. See M. Boretsky, "Concerns about the Present American Position in International Trade," *Symposium on Technology and International Trade* October 14-15, 1970 (Washington, D.C.: National Academy of Engineering, 1971); M. Boretsky, *The Threat to U.S. High Technology Industries: Economic and National Security Implications* (Washington, D.C.:

high- from low-technology industries and products. Regardless of the measure used, all three techniques are in rough agreement as to which industries and products can be considered high technology.

In terms of North-South technology transfer, all of the products and industries found to be technology intensive by any of the techniques mentioned above can usefully be included in a definition of high technology. In this way, some high-technology subsectors of low-technology industries are included, as well as some product groups (such as automobiles and radio and television receivers) that would not be considered as high technology in the context of technology transfer between IACs. Industries and products where "soft" technology, such as marketing and management know-how, are important sources of competitive strength (and essential elements of a transfer) should also be included.

High technology refers to a sophisticated range of products (from automotive equipment to biogenetic materials), components (from diesel engines to semiconductors), management (from quality circles to informatics), and production techniques (implicit in the quantum and complexity of industrial know-how). Any definition of high technology is necessarily relative and depends in part upon the technological gap between technology suppliers and receivers: the formers' ability to impart technical know-how (packaged or otherwise) effectively in relation to the technical absorptive capacity (at the enterprise level) of the latter.

U.S. Department of Commerce, International Trade Administration, 1982). In order to account for the wide range of products produced by any one industry, Kelly developed a product-based index using applied R & D expenditures by product field and the value of product shipments to devise technology-intensity ratios. See R. K. Kelly, "The Impact of Technological Innovation on International Trade Patterns," Staff Economic Report (Washington, D.C.: U.S. Department of Commerce, Office of Economic Research, 1977). Aho and Rosen used essentially the same technique as Kelly with more recent R & D expenditures data and shipments data on a product line basis. See C. M. Aho and H. F. Rosen, "Trends in Technology-Intensive Trade," Economic Discussion Paper No. 9 (Washington, D.C.: U.S. Department of Labor, Bureau of International Labor Affairs, 1980). Davis developed a technique using input-output analysis to measure total R & D (R & D contributed by inputs plus R & D expenditures directly on the product in question). Product groups were then ranked by their total R & D to shipments ratio on an index of technology intensity. See L. A. Davis, *Technology Intensity of U.S. Output and Trade* (Washington, D.C.: U.S. Department of Commerce, International Trade Administration, 1982).

A distinction should be drawn between technology transfers that impart operational, duplicative, and innovative capabilities.* Implantations of operational (turnkey) technology theoretically permit the recipient to make a product equivalent to that produced by the technology supplier, though in practice the production efficiencies and quality levels achieved vary widely. Duplicative technology transfer implies design and engineering know-how sufficient to reproduce an entire plant or discrete components. Technology transfers that engender innovative capability allow the recipient to go beyond duplication to alter transferred products, processes, and equipment designs in response to changing resources, requirements, and market demands. None of these capabilities will accrue as a matter of course to passive recipients of technology; each requires an increasing level of technological effort.

High technology, as defined above, is being transferred to a wide range of developing countries. Mansfield notes the increasing overall rate of international technology transfer and points to the growing role of multinational corporations (MNCs) as transfer agents.[1] A great deal of soft technology is transferred in ways that do not involve payment (for example, through personnel exchanges, publications, conferences, exhibitions, foreign study, and so on) and are difficult to quantify. Several studies have examined the relative importance of different modes of transferring both high and low technology in several countries.[2]

The highest technology, however, is concentrated in a relatively small number of firms (mostly MNCs) that are based in the IACs. Because of this concentration and the growing need for technological upgrading as an input to economic growth, most North-South transfer of high technology will continue to be made on a commercial basis. While several indicators have been used to monitor the international transfer of technology,[3] they must be refined to reflect better: 1) what has actually been transferred (that is, the quantum

---

*Dahlman and Westphal draw roughly the same distinctions, using a different terminology, to describe the broad categories of activity in which technological knowledge is applied within industry: 1) production engineering, 2) project execution, 3) capital goods manufacturing, and 4) research and development. They note that technological mastery is usually acquired in these activities in the order shown above. See C. J. Dahlman and L. E. Westphal, "The Meaning of Technological Mastery in Relation to Transfer of Technology," World Bank Reprint Series no. 217 (Washington, D.C.: World Bank, 1982).

and complexity of the technology involved; 2) the full market value price of technology actually transferred (using estimates of total revenues received by suppliers, and of any monopoly rents); and 3) the utility of received technology to the user, involving estimates of value that can be translated into market competitiveness once the technology is absorbed.[4]

For the reasons cited above, accurate measurement of the transfer of high technology is especially difficult, given the constantly evolving relationships between suppliers and receivers of technology (see below). In view of these difficulties, the case study method of analyzing technology transfer[5] will continue to provide critical insights and information necessary for a better reading of trends revealed by gross indicators of high-technology transfer.

TRENDS IN NIC AND DEVELOPING COUNTRY DEMAND FOR HIGH TECHNOLOGY

Not all countries in the South are eager to receive high technology transferred from IACs. Developing countries whose patterns of consumption approach those of the developed countries and whose main trading orientation is with the IACs have a greater need to use high technology in their modern sectors. This group of countries includes primarily the NICs in Latin America (for example, Brazil, Mexico, and Argentina) and in Asia (for example, Hong Kong, the Republic of Korea, Singapore, and Taiwan). Individual enterprises in many of these NICs have acquired considerable operational expertise, as well as the ability to duplicate transferred technology.

Other developing countries, with much lower capability to absorb technology, are also seeding transfers of high technology, in spite of the high costs and risks involved in terms of investment capital and socioeconomic development goals. This is particularly true of oil-producing countries flush with foreign exchange earnings. Saudi Arabia, for example, has to date received tens of billions of dollars worth of high technology (mainly petrochemicals), primarily through the mechanism of joint ventures with firms from the United States, Europe, and Japan and at least one firm from Taiwan.[6]

Trends in NICs and developing countries that have contributed to the greater emphasis on absorption of high technology are 1) upward movement by the South in the international division of labor,

2) a shift from import substitution to export promotion strategies, and 3) the emerging role of national policies for technology-intensive industrial development.

Upward Movement in the International Division of Labor

For several years there has been mounting pressure on both developing countries and NICs to move to ever higher levels of technology. The Taiwan Fertilizer Company's joint venture (50 percent equity) with Saudi Arabia is indicative of the shift toward high technology by the South. Taiwan has achieved sufficient operational and duplicative capability in fertilizer production to be chosen as a technology supplier and investor by Saudi Arabia, which had severe problems in the operation of a turnkey fertilizer plant built in the early 1970s by the Occidental Petroleum Corporation.[7]

As Saudi Arabia gains operational capability, its comparative advantage in fertilizer production (based largely on huge natural gas resources) increases while that of Taiwan declines. Saudi adoption of Taiwanese fertilizer technology thus exerts pressure on Taiwan to seek comparative advantage in higher technology. Other developing countries are following this trend and building fertilizer plants under a variety of similar technology transfer arrangements.

This process can be seen at work in many mature industries such as textiles, steel, and black-and-white television, as the product life cycle works its way through to rising developing countries such as Malaysia, the Philippines, and the People's Republic of China. The technology these countries seek to receive is high in relation to their level of technological mastery, and its successful implantation forces the NICs to seek comparative advantage in higher levels of technology, which in turn exerts pressure on the IACs to move on.

The Shift from Import Substitution to Export Promotion

Many of the newly industrializing countries of the South are in the process of moving from import-substituting industrial development to export promotion strategies. These nations are attempting to pursue the Japanese model of technological development, with mixed results. Japanese industry did not suddenly develop the ability to reap the advantages associated with vigorous export market penetration. Even though export specialization was pursued as a deliberate goal early on in Japan, policy makers realized that a prolonged period of protectionism (including exclusion of foreign goods and foreign direct investment wherever possible) would be necessary to build a strong home market before domestic industry could reach internationally competitive production volumes and otherwise become effective in exporting.

Japan pursued a "search and bring back" policy with regard to foreign technology, relying heavily on its own rapidly developing human resources to forge indigenous technological development. The success of this policy was due in part to the presence of a residual pool of technical personnel after World War II. Yet Japan's primary advantage lay in its singular instinct for sensing the appropriate time to change its industrial orientation, and in then pursuing appropriate policies to encourage the desired change.

The shift to export promotion on the part of the NICs has not followed as deliberate nor as well-timed a course as that of Japan. The NICs did not have a pool of technical personnel to draw on during import substitution and so relied on the process of technology transfer to increase the level of local technological mastery. Many NICs also have high levels of foreign direct investment, giving their modern sectors a strong trade orientation toward the IACs.

This trade orientation and the need to earn foreign exchange to service foreign debt have led many NICs to devote increasing attention to export promotion, a shift that is occurring, in many cases, before significant indigenous technological capability has been developed. The growing importance of technological change in maintaining international competitiveness requires NICs to seek ever higher levels of technology if export promotion strategies are to succeed.[8]

The Role of National Policies in Technological Development

Many (if not most) countries in the South have the objective of upgrading indigenous technological capabilities through technology transfer. A related (and often conflicting) goal is to reduce dependence on the IACs as sources of technology. The overriding objective of most developing country policies regarding technology transfer, however, has been to reduce the cost of imported technology by regulating the form of technology inflows, by avoiding duplication, and by regulating the conduct of technology suppliers.[9] Appropriate policies vary according to the stage of development of the country, its administrative and technological capability, and the particular industry included.

Many of the NICs, having gained considerable operational and duplicative capabilities through earlier technology transfers, are now trying to encourage development of innovative capabilities. This often requires the transfer of certain high technologies that are considered tools of innovation needed to compete in modern global markets (for example, machine tools, computers, automated manufacturing systems, and telecommunications). It also requires the formation of technical personnel who can "penetrate the logic" of acquired technology (see below).

India, for example, has sought to develop a domestic machine tool industry, considering it a sine qua non for technological autarky. As a highly protected economy, India pursued this capability through judicious licensing from suppliers in the IACs. Over the course of 30 years, India's Hindustan Machine Tools, Ltd. (a public sector company) has produced machine tools of ever-increasing sophistication and diversification, following the Western pattern of competition based on product innovation.[10] By the late 1970s innovative capabilities had progressed to the point where the company could design and build computer numerically controlled machining centers.

A crucial aspect of innovation is the product marketability that should result from the adaptation and modification of transferred technology. Japanese firms realize that product marketability must be explicitly linked to a growing innovative capability in order to compete at home and abroad. In this sense, India's innovative capability in machine tools remains just that – an unrealized capability – since its machine tool industry produced only two computer numerically controlled machining centers in 1980 (compared to the 4,280 units produced the same year in Japan). The centers, developed without competitive pressure in India's protected economy, are too costly for either domestic or export buyers.

While India suffers from an overly protected economy in its search for technological mastery, other NICs have been able to utilize increasing national technological capability to support (in a scientific manner) the development of a coherent industrial policy, thus moving beyond primarily defensive administrative controls on technology transfer. Brazil, for example, has targeted telematics (the combination of telecommunications and electronic information processing, or informatics) as an important industrial sector for the future, wherein the transfer of high technology should be encouraged.

Brazil's special secretariat for informatics administers a policy in which the more sophisticated products may be produced by MNCs, which are encouraged to export and to improve their local R & D facilities.[11] As soon as a product can be manufactured by domestic firms, the corresponding market segments are protected. When domestic firms become internationally competitive, protection is removed. This policy is intended to shift Brazil's comparative advantage toward more innovation- and technology-intensive industrial sectors by encouraging domestic industry to manufacture increasingly complex products and technologies.

IAC RESPONSES TO INCREASING DEMAND FOR HIGH TECHNOLOGY

Suppliers of high technology (MNCs in large part) have traditionally been reluctant to transfer what they consider "core" technology (that is, unique or proprietary know-how that is the basis of a competitive advantage) to noncontrolled firms overseas.[12] In spite of the increasing pace of technology transfer, the mean age of technology transferred by U.S.-based MNCs to noncontrolled foreign firms continues to be significantly higher than that of technology transferred to subsidiaries in both the IACs and developing countries.[13] This is one indication that core technology is not "leaking out" to the detriment of industry in the IACs.

Our definition of high technology includes, but is not limited to, such core technology. High technology will continue to be transferred to the South in a variety of ways and for several reasons: 1) the new relationships between MNCs and developing country enterprises; 2) continuing MNC efforts to optimize global operations; 3) MNC efforts to maintain market presence in expanding developing country/NIC markets; and 4) the desire of IAC enterprise to earn returns on "standardized" technology.

New Relationships between MNCs and Developing Country Enterprises

Until the end of the 1960s, the establishment of the wholly owned foreign subsidiary or the majority-owned foreign affiliate was the predominant method of foreign expansion by MNCs and a prime source of technology transfer. Data on foreign direct investment were taken as the main, if not the only, quantitative indicator of such expansion. By the late 1970s it became clear that joint ventures had become far more important as a vehicle of overseas investment by MNCs.[14]

Moreover, other new forms of interface between MNCs and developing countries are emerging and spreading. The direct investment package is rapidly giving way to complex joint-venture agreements, production-sharing agreements, management and marketing contracts, service agreements, technology licensing contracts, turnkey and similar contracts, countertrading arrangements, and numerous other forms of nonequity interaction. While the coverage of the available data on these phenomena is not as wide as that on foreign direct investment, it seems clear that a marked rise in the relative importance of such new relationships is occurring.

This growing shift of a fundamental qualitative nature in the principal ways in which MNCs engage in transactions in the world economy is in part a response to a number of structural shifts and policy changes in the international environment in which these enterprises operate. Following the wave of nationalizations in the petroleum and extractive industries in the early 1970s, many of the newly created state-owned enterprises found it expedient to negotiate contracts for the provision of management, marketing, and technical services with their former MNC owners, so as not to disrupt production and marketing flows and to maintain access to foreign technology.

The growing bargaining power of domestic enterprises in many developing countries, especially the larger and more technically advanced NICs, has increased pressure for the "unbundling" of the technology, management and technical services, and financing normally integrated in the foreign direct investment package. Such pressure is frequently reinforced by government policies excluding foreign direct investment from certain sectors and insistence on joint ventures or limited foreign ownership in others. This type of environment for technology transfer is also prevalent in countries where public sector companies have a leading role in economic development.

It is not necessarily true that the bargaining power of developing country enterprises (and governments) is decreased in proportion to the sophistication of the technology sought. A better indicator of the relative bargaining strengths of supplier and purchaser might be the dynamic (as opposed to stable) nature of the technology in question.[15] It is more difficult for NICs to bargain for and master high technology that is undergoing continuous change in production processes or in products (such as semiconductors). The introduction of automated manufacturing systems (which permit rapid change in product and process) by the IACs may widen the gap between technology suppliers and developing country enterprises in a variety of industries.

A third important change in the operating environment for MNCs is the growth of international lending by multinational banks. Their willingness to undertake vastly greater developing country debt has effectively increased the opportunities for providing the capital and foreign exchange components independently of the other elements in the foreign investment package and in that sense has facilitated the spread of nonequity arrangements.

The new technology-sharing agreements represent a radical departure from the traditional approach U.S. corporations have taken toward managing technological assets. In the past it was only late in a product's life cycle – once the production techniques had become generally available – that the corporation was willing to release technology. Case study materials reveal that a growing segment of industry in the IACs is now prepared to transfer high technology under terms that assure rapid and efficient implantation of an internationally competitive production capability in the host country.[16]

While the stimuli to do so may have been external, MNCs have rapidly adopted nonequity relationships as internal parts of their overseas strategies. Not only in situations where direct investment is excluded, but also where equity investment is perceived to be too risky or where the cost of capital is high, such arrangements represent alternative means of earning returns on the firm's technological, management, or marketing assets. Moreover, such arrangements often give the MNC a degree of operational control over the host enterprise without any formal equity involvement whatever.

MNC Efforts to Optimize Global Operations

The organization and logistics of world production, marketing, and the design and engineering of complete industrial systems is undergoing deep-seated change. Products are designed and production processes are engineered to accommodate not only the multiplicity of markets they must serve but also their specialized manufacture and interchange among both affiliated and sovereign enterprises. This new international division of labor requires that MNCs optimize their global operations in several ways that imply continuing transfers of high technology to the South.

The large automakers, for example, may support the transfer of automated manufacturing systems (robotics) to overseas subsidiaries and affiliates engaged in the production of components for the new generation of "world cars." This possibility is enhanced by the need to meet international standards of quality and cost. A growing proportion of R & D expenditure by MNCs is located in developing country laboratories. This is partly due to the enhanced bargaining power of the developing countries but also is part of global optimization, as MNCs seek to link R & D better to local markets and to amortize R & D costs over revenues derived from many countries.

MNC Efforts to Gain or Maintain Market Presence

Some willingness to transfer high technology in the South results from the necessity to gain or maintain a presence in expanding markets considered vital for the future. Increased bargaining power and ability to regulate the conduct of MNCs enables many developing countries to demand the transfer of desired high technology in return for the right to operate in the country. Some high technology may be willingly transferred to local component suppliers, for example, in order to maintain quality and cost standards where local content laws exist.

At other times, high technology is transferred unwittingly. Saudi Arabia, for example, spent seven years researching and studying detailed bids by several MNCs for a sophisticated telecommunications network.[17] Thus, their desire to maintain a presence in the lucrative Middle East market required that these MNCs transfer a significant amount of know-how to the Saudi R & D evaluators at very low cost.

The Desire to Earn Returns on Standardized Technology

Many MNCs transfer high technology as a result of decisions to earn returns on technological assets that are not considered essential to the company's business, that are widely available elsewhere, or that are no longer commercially viable. In this sense, the particular high technology sought by a developing country enterprise is important. Much petrochemical process technology, for example, is widely available on relatively easy terms (that is, much of the technology is nonproprietary, and other key processes are widely licensed by both technology originators and international contractors). In addition, many petrochemical processes can be duplicated with slight modifications that make it difficult to prove patent infringement. This proliferation in the sources of some high technology is an incentive for technology suppliers in the IACs to transfer it in order to gain whatever return is possible.

CURRENT DEFICIENCIES, PROBLEMS, AND ISSUES

High technology flows to the more advanced sectors of NICs will depend on the relative strengths of pull factors on the demand side, coupled with push factors on the supply side. On the demand side, the major determinants are: 1) commercial astuteness and technical absorptive capabilities at the enterprise level; and 2) public policies to promote the transfer and absorption of high technology, particularly financial mechanisms to mobilize and channel required resources within a particular country. On the supply side in the IACs, major determinants include: 3) the nature and direction of the new automation; and 4) industrial policies, including labor market adjustment mechanisms that are supportive of outward flows of high technology.

Commercial and Technical Factors in the NICs

In the acquisition, implantation, and commercial utilization of high technology, much depends upon enterprise capabilities to develop both domestic and foreign markets and to adapt received technology to the technical absorptive capabilities of both the recipient enterprise and the component supplier industries. Even countries such as Brazil and South Korea, with substantial industrial infrastructure, still have very significant gaps in the kind of technical personnel needed to adapt and manipulate technology at anywhere near the level of capabilities found in the IACs (Japan in particular).

In South Korea, for example, enough technical people are available to derive both operative and duplicative capabilities from transferred technology. South Korea is at a serious disadvantage, however, when it comes to the adaptive or innovative stage, where it is necessary to penetrate the logic of received technology and to adjust scale, component design, raw material inputs, or machine performance characteristics of manufactured products.[18]

Public Policies and Financial Mechanisms in the NICs

In countries like South Korea, Brazil, and Mexico, comprehensive policies and measures have been developed to promote, enhance, and support high-technology flows.[19] In each of these countries, substantial efforts have been made to develop special financial mechanisms to provide the required foreign exchange resources and to assist indigenous enterprise in the acquisition of technology from abroad.[20]

One of the more advanced efforts in this direction is the Korean Technology Development Corporation, which was organized in 1981 with the support of a World Bank loan, to finance software-intensive R & D activities by South Korean enterprises.[21] The founding of the corporation was in response to the recognized need to upgrade technical absorptive capabilities, particularly of those Korean firms anxious to enter the more competitive international markets.

Impact of Automated Manufacturing Systems in the IACs

The rapid movement in Japan, the United States, and Western Europe toward automated manufacturing systems, including computer-controlled machines, integrated machining centers, robotics, and computer-aided design and engineering, is bound to widen even further the existing gaps in high-technology areas.[22] In all likelihood it will also reduce (in the short run at least) the range of industrial activities that will be transferred to developing countries and NICs, which are now used by some MNCs as offshore manufacturing facilities for world markets.

There are several reasons for this trend. In the first place, these new systems require (and at the same time permit) high orders of precision and meticulous quality control standards. Secondly, component manufacturers will have to be more closely integrated (in terms of proximity and technical interaction) in order to realize the benefits of tighter production inventory control, total quality control, and the entire range of integrated design, engineering, and production function the new automation makes possible. Thirdly, automated manufacturing systems with quickly reprogrammable component designs and tooling permit efficient manufacture of much smaller production volumes than was previously the case with inflexible, high-volume, dedicated production equipment.

In the past, offshore manufacturing in developing countries has proven cost-effective because of significantly lower labor costs, which often offset (at least in part) learning costs and the diseconomies of lower production volumes. Even as wage rates rise in the developing countries, this advantage is now being narrowed by automated manufacturing systems that are cost-effective even at low volumes, and the transmigration and learning gaps are considerably enlarged.

Industrial Policies in the IACs

The pace and direction of change in the North-South division of labor in high-technology industries will depend heavily upon the industrial policies adopted in the countries of the developed world. Japan has a coordinated set of policies to shift its industrial labor force into high-technology "sunrise" industries and to phase out declining "sunset" industries. This shift is commensurate with the increasing knowledge and skills of Japan's work force and with other population changes. Indirectly, the Japanese are consciously making space for the emerging newly industrializing countries (NICs). Japan's relatively high growth rates, low unemployment, and effective labor-market adjustment mechanisms facilitate this process of transition.

The situation is very different in the United States and parts of Western Europe, where depressed growth rates, high unemployment, and protection of declining industries inhibit and retard the process of transition and adjustment.[23] The rate at which IACs move into automated manufacturing and are willing to spin off declining industries (which still represent high technology in the eyes of the developing countries) depends upon these differences in industrial policy and economic management. The current trend toward protectionism in some IACs will inevitably slow down North-South transfers of high technology.

NOTES

1. E. Mansfield, "International Technology Transfer: Rates, Benefits, Costs, and Public Policy" (Philadelphia: University of Pennsylvania, 1981): pp. 16-20.
2. T. H. Moran, "Corporate Bargaining Strength and Strategies to Offset Political Uncertainty," chapter 12 in *International Political Risk Assessment: The State of the Art* (Washington, D.C.: Georgetown University, School of Foreign Service, 1981): pp. 11-18.
3. J. Baranson, "Technology Licensing Guidelines for U.S. Business in India." (Washington, D.C.: Developing World Industry and Technology, 1981).
4. J. Baranson and H. Malmgren, "Technology and Trade Policy Issues: An Agenda for Action" (Washington, D.C.: Developing World Industry and Technology, 1981).
5. J. Baranson, *Technology and the Multinationals* (Lexington, Mass.: Lexington Books, 1978): pp. 11-14.
6. J. Baranson and R. Roark, "How to Compete in Middle East Technology Markets: Telecommunications Systems and Petrochemical Plants" (Washington, D.C.: Developing World Industry and Technology, 1983): p. 37.
7. Ibid., p. 36.
8. Baranson and Malmgren, "Technology and Trade Policy Issues."
9. T. Sagafi-nejad, R. W. Moxon, M. V. Perlmutter, eds., *Controlling International Technology Transfer: Issues, Perspectives, and Policy Implications* (New York: Pergamon Press, 1981); F. Stewart, "International Technology

Transfer: Issues and Policy Options" (Washington, D.C.: World Bank, 1979): pp. 46-77.

10. R. Matthews, "Industrial Strategy and Technological Dynamism in Machine Tool Manufacture: Comparative Perspectives on India and Japan," *Technology and Culture*: Occasional Report Series no. 7 (Lund, Sweden: University of Lund, Research Policy Institute, 1983): pp. 10-14.

11. J. de O. Brizida, "Transborder Data Flows in Brazil," in *The CTC Reporter*, no. 13 (New York: United Nations, Centre on Transnational Corporations, 1982): p. 12.

12. Baranson, *Technology and the Multinationals*, p. 12.

13. T. S. Kim, "The Korea Technology Development Corporation," *Policy Choices* (Cambridge, Mass.: Center for Policy Alternatives, Massachusetts Institute of Technology, Fall 1983): p. 2.

14. Baranson, *Technology and the Multinationals*, p. 10.

15. Moran, "Corporate Bargaining Strength," p. 77.

16. Baranson, *Technology and the Multinationals*; J. Baranson, *North-South Technology Transfer: Financing and Institution Building* (Mt. Airy, Md.: Lomond Publications, 1981).

17. Baranson and Roark, "How to Compete in Middle East Technology Markets," pp. 21-29.

18. Baranson and Malmgren, "Technology and Trade Policy Issues," pp. 140-41, 144-45.

19. Ibid., pp. 112-53.

20. J. Baranson, *North South Technology Transfer*, pp. 45-47, 94-98.

21. T. S. Kim, "The Korea Technology Development Corporation," pp. 4-5.

22. J. Baranson, "Automated Manufacturing: The Key to International Competitiveness and Why the U.S. Is Falling Behind" (Washington, D.C.: Developing World Industry and Technology, 1983): pp. 1-5.

23. Ibid., pp. 13-14.

REFERENCES

C. M. Aho and H. F. Rosen. "Trends in Technology-Intensive Trade." Economic Discussion Paper no. 9. Washington, D.C.: U.S. Department of Labor, Bureau of International Labor Affairs, 1980.

J. Baranson. "Automated Manufacturing: The Key to International Competitiveness and Why the U.S. Is Falling Behind." Washington, D.C.: Developing World Industry and Technology, 1983.

——. "Changes in the Terms and Conditions of Technology Transfer by the Pharmaceutical Industry to Newly Industrializing Nations: An Overview of the Past Ten Years." Washington, D.C.: Developing World Industry and Technology, 1979.

——. "Critique of International Technology Transfer Indicators." Background Paper for *Science Indicators 1980, Vol. 1: Indicators of International Technology and Trade Flows*. Washington, D.C.: National Science Foundation, 1980.

——. "Mechanisms to Finance Sales of Technology Support Services by U.S. Enterprises to Newly Industrializing Nations." Washington, D.C.: Developing World Industry and Technology, 1980.

——. *North-South Technology Transfer: Financing and Institution Building.* Mt. Airy, Md.: Lomond Publications, 1981.

——. *Technology and the Multinationals.* Lexington, Mass.: Lexington Books, 1978.

——. "Technology Development Project: Field Mission to the Republic of Korea." Washington, D.C.: Developing World Industry and Technology, 1979.

——. "Technology Licensing Guidelines for U.S. Business in India." Washington, D.C.: Developing World Industry and Technology, 1981.

J. Baranson and H. Malmgren. "Technology and Trade Policy Issues: An Agenda for Action." Washington, D.C.: Developing World Industry and Technology, 1981.

J. Baranson and R. Roark. "How to Compete in Middle East Technology Markets: Telecommunications Systems and Petrochemical Plants." Washington, D.C.: Developing World Industry and Technology, 1983.

M. Boretsky. "Concerns about the Present American Position in International Trade." Report. In *Technology and International Trade.* Washington, D.C.: National Academy of Sciences, 1971.

——. *The Threat to U.S. High Technology Industries: Economic and National Security Implications.* Washington, D.C.: U.S. Department of Commerce, International Trade Administration, 1982.

J. de O. Brizida. "Transborder Data Flows in Brazil," in *The CIC Reporter,* no. 13. New York: United Nations, Centre on Transnational Corporations, 1982.

C. J. Dahlman and L. E. Westphal. "The Meaning of Technological Mastery in Relation to Transfer of Technology." World Bank Reprint Series no. 217. Washington, D.C.: World Bank, 1982.

L. A. Davis. *Technology Intensity of U.S. Output and Trade.* Washington, D.C.: U.S. Department of Commerce, International Trade Administration, 1982.

R. K. Kelly. "The Impact of Technological Innovation on International Trade Patterns." Staff Economic Report. Washington, D.C.: U.S. Department of Commerce, Office of Economic Research, 1977.

T. S. Kim. "The Korea Technology Development Corporation." In *Policy Choices.* Newsletter. Cambridge, Mass.: Center for Policy Alternatives, Massachusetts Institute of Technology, Fall, 1983.

E. Mansfield. "International Technology Transfer: Rates, Benefits, Costs, and Public Policy." Philadelphia: University of Pennsylvania, 1981.

R. Matthews. "Industrial Strategy and Technological Dynamism in Machine Tool Manufacture: Comparative Perspectives on India and Japan." *Technology and Culture*: Occasional Report Series no. 7. (Lund, Sweden: University of Lund, Research Policy Institute, 1983.

T. H. Moran. "Corporate Bargaining Strength and Strategies to Offset Political Uncertainty." Chapter 12 in *International Political Risk Assessment: The State of the Art.* Washington, D.C.: Georgetown University, School of Foreign Service, 1981.

S. W. Rosoom. "The International Technology Transfer Process." Washington, D.C.: National Academy of Sciences, 1980.

T. Sagafi-nejad, R. W. Moxon, M. V. Perlmutter, eds. *Controlling International Technology Transfer: Issues, Perspectives, and Policy Implications*. New York: Pergamon Press, 1981.

F. Stewart. "International Technology Transfer: Issues and Policy Options." Washington, D.C.: World Bank, 1979.

Part II

Historical Experiences of Individual Countries

3

Trade in Technology by a Slowly Industrializing Country: India

Sanjaya Lall

INTRODUCTION

The analysis of technology transfer from the industrialized to the developing countries has passed through several phases.[1] It started with the problems of implementing modern production methods in relatively primitive environments and the costs of transfer in inherently imperfect and oligopolistic markets for knowledge. It then moved on to problems of the socioeconomic appropriateness of imported technologies and their adaptation to developing country needs. More recently, it has concerned itself with the interaction of technological generation and technology imports in the newly industrializing countries (NICs). One strand of this literature, in which I have been particularly involved, has concentrated on exports of technology by (including the emergence of multinational enterprises from) the NICs.[2]

While the substantial accumulation of empirical literature on different aspects of the technology transfer process has enriched our understanding, we are still far from a comprehensive analysis that brings together all the diverse economic determinants of technology import, assimilation, adaptation, innovation and export. Clearly

---

This paper draws upon material collected as part of a project sponsored by the World Bank. I wish to acknowledge my gratitude to the bank, and to take upon myself the responsibility for the views expressed here. I am also grateful to Rajir Kumar and Ritu Kumar for their help in collecting the data.

these all comprise dynamically interlinked elements of a complex process. The fact that most NICs are now simultaneously importing and exporting technologies suggests that there are economic variables that determine their dynamic comparative advantage in the "production" of technology. More interestingly, the fact that different NICs have different patterns of "revealed" comparative advantages in their technology trade suggests that special attention be directed at *differences* in their endowments and strategies. These differences, apart from the obvious ones of market size, rate of growth, stocks of scientific manpower, and the like, relate to a wide array of official policies: industrial promotion (in particular, of the capital goods sector, which Rosenberg[3] rightly regards as the hub of technical progress), protection of domestic production against imports, entry of foreign technologies in various forms, inducements to local technological activity, and the broader set of regulatory policies that make for greater (or lesser) efficiency in industrial production.

Many of these factors have received preliminary examination in the budding literature on Third World technology,[4] though not generally with the purpose of contrasting the effects of different policy regimes among the NICs. This chapter, while concentrating on India's experience as technology trader, will touch upon some of these contrasts.

The Indian case has several points of interest. Despite its low income levels, the country possesses a relatively large pool of skilled technological manpower. Its industrial performance has been, on average, rather poor in comparison to other NICs, but it has built up one of the most diverse productive bases in the Third World, particularly in capital goods manufacture. Its obsessively inward-looking strategy has saddled it with large areas of inefficiency (and has resulted in abysmal export performance), yet it has emerged as one of the leading Third World exporters of industrial technology. This indication of diverse and in-depth technological development coexists with a relatively low reliance on imports of modern technology and a fairly widespread incidence of technological obsolescence in industry.

There are, at least at first sight, several paradoxes in the Indian case. These cannot be resolved with the evidence at hand, but this chapter will present my own interpretation of the technological issues involved. I need hardly emphasize the tentative nature of the arguments.

First I briefly review the evidence on India's imports and exports of technology. Using the conventional terminology of the transfer of technology literature, technology trade is taken to exclude transactions in which only physical products are involved and also all nonmarket technology transactions, as well as the migration of technical personnel. Furthermore, I deal only with industrial technology; this ignores the export of civil construction services by India, quantitatively more important than industrial technology exports but of minor interest in the present context.

My focus is, therefore, on the following forms of technology imports (TI) and technology exports (TE): industrial turnkey projects; direct foreign investment; consultancy services; and sale of know-how, patents, trademarks, and so on in the form of one-off deals or continuous contractual agreements (loosely labeled "licensing" for convenience).

I then provide an interpretation of Indian performance, after which I sum up the main conclusions.

TECHNOLOGY IMPORTS BY INDIA[5]

Let us start with some background, for those who are unfamiliar with Indian industry. Since the late 1950s India has pursued an industrial strategy dominated by the following objectives: conservation of foreign exchange and, related to this, the domestic production of everything that it was physically possible to produce (regardless of cost), with priority laid on heavy capital and intermediate goods; the building up of a scientific/technological infrastructure and the development of self-reliance in industrial technology; the national ownership of industrial enterprises, with the commanding heights reserved for the public sector; and the encouragement of small enterprises together with the containment of large private enterprises. These various objectives, sometimes closely intertwined and sometimes conflicting, have been pursued by the setting up of an administrative apparatus of incredible complexity, which has sought, at the same time, to accommodate populist (and self-seeking) political pressures, to decentralize industry, to promote a largely incompetent public sector, to implement some of the highest tax rates in the world, and to define, by arbitrary administrative fiat, what a "socialist pattern of society" meant in a highly unequal and mainly market-oriented economy.[6]

It would be well beyond my present purposes to describe these general policies in greater detail, though some of them are relevant to the discussion and will be resurrected. The result has been that, after a respectable head start in the 1950s, India has experienced highly erratic but on average declining industrial growth rates in the 1960s and 1970s. Over these two decades its individual progress has been the slowest of all developing countries with respectable industrial sectors with the sole exception of Argentina.

In 1979, according to the *World Development Report* (*WDR*) 1982, India's manufacturing value added came to $15.6 billion (in 1975 dollars),* as compared to $40.3 billion for Brazil, $23.4 billion for Mexico, $11.2 billion for Argentina, $10 billion for South Korea, $12.8 billion for Yugoslavia, $3.6 billion for Hong Kong, and $2.1 billion for Singapore.† Within manufacturing, the value of machinery and transport equipment production came to $3.0 billion for India compared to $11.3 billion for Brazil, $4.2 billion for Mexico, $2.7 billion for Yugoslavia, $2.5 billion for Argentina, $1.9 billion for South Korea, and $1.0 billion for Singapore. This may be used as an indicator of capital goods production capability, but the inclusion of passenger cars in the total tends to inflate the figures for the Latin American countries relative to the others. If we allowed for this, I would guess that the ranking of capital goods manufacturers in the Third World would be roughly: Brazil, India, Yugoslavia, Mexico, Argentina, South Korea, and Singapore (with Taiwan just behind South Korea).

We can evaluate Indian TI against this setting. Given the value of its industrial production, India seems to have the lowest relative reliance on foreign technology of all the NICs in the past 15 years or so. Let us take the data for each form of TI in turn.

The major form of technology transfer in most developing countries is direct foreign investment. India had inherited, from colonial times, a relatively large stock of overseas investment. It

---

*India ranks fifth in the developing world, after Brazil, China, Israel, and Mexico, and around twenty-third in the world as a whole (including Eastern Europe and Switzerland). Exact rankings are not possible because the *WDR* does not provide industrial value added data for Switzerland or the Council for Mutual Economic Aid (COMECON) countries.

†Data for Taiwan are not given in *WDR*, 1982, but in 1976 its manufacturing value added was just ahead of South Korea's. It may be placed at $11-12 billion for 1979.

pursued a liberal policy toward foreign entry until the early 1960s, when the first phase of its import-substituting industrialization attracted a number of foreign industrial investors in sectors permitted to them. By 1967, its stock of direct investment came to $1.3 billion, compared to $3.7 billion in Brazil, $1.9 billion in Argentina, $1.8 billion in Mexico, and $78 million in South Korea. From 1968 onward, India applied increasingly restrictive policies to foreign capital inflows and, from 1974 on, to the internal expansion of majority-controlled foreign subsidiaries. It forced most foreign affiliates to dilute their shareholding to 40 percent; some 60 companies decided to close their operations in India. Fresh applications for foreign investment were subjected to lengthy, cumbersome, and stringent scrutiny by the government, which sought to let multinational corporations (MNCs) only into what were classified as high-technology activities where domestic capabilities were inadequate, or into export-oriented activities.

Not surprisingly, in the period 1969-82 India approved fresh equity inflows of only $80 million in total. Of these approvals, less than half materialized, so that only about $40 million actually entered the country in the past 14 years. Compare this with a *net* inflow (net of repatriations) of $14 billion in 1970-80 in Brazil, $7 billion in Mexico, $648 million in South Korea (fresh new investment in South Korea from abroad came to $1.2 billion in 1967-81[7]), and even $1.5 billion in relatively stagnant Argentina (Taiwan probably took about the same amount of foreign capital as South Korea in this period). In terms of *net* inflows, India had a substantial negative total for the 1970s.

As far as imports of foreign turnkey projects and the use of foreign consultants are concerned, no data on values are available for India or any of the other NICs. We know, however, that in the period 1957-67, 254 agreements for plant construction (mostly turnkey) were signed with foreign firms by India, but since then a rule has been passed that only Indian firms could act as prime consultants or contractors. As a result, there was no foreign turnkey work given to foreign firms until the 1980s. It is only in the past two years that the government has permitted some large fertilizer and aluminum contracts (financed from abroad) to be awarded abroad. Thus, for much of its recent industrial history, India has relied on its domestic consultants, of whom only a few have direct equity links abroad. These consultants have licensed foreign processes and drawn upon some specialized foreign technical services, of course, but by and large they have attained a high degree of competence in implementing and adapting technologies from the basic design stages. I would hazard a guess that in the 1970s India had the least reliance on foreign consultants and contractors of all the NICs.

Data on licensing payments abroad are slightly easier to come by. It may be thought that India would, by eschewing a reliance on direct participation by MNCs and foreign consultants, tend to rely relatively more heavily on foreign licensing. And, indeed, licensing has been the favored means of access to technologies not available locally (or considered by the officials not to be so available). In the period 1969-82, the government approved 3,323 technical collaboration agreements (not involving any equity participation), with the largest proportion coming from the United States and the United Kingdom, followed by West Germany, Japan, France, and Italy. The entire sum of foreign collaborations since 1957 (this includes both direct investments and licensing) comes to 6,959, with the following industrial distribution:

Industrial machinery and machine tools	2,754 (40%)
Electrical equipment and electronics	1,293 (19%)
Chemicals and pharmaceuticals	1,056 (15%)
Transport and construction equipment	599 (9%)
Technical consultancy	101 (1%)
Other	1,156 (16%)
Total	6,959

Source: Indian Investment Centre, 1982, p. 17.

The marked emphasis of technology import policy on capital goods and advanced intermediates (chemicals) is obvious from the distribution of approvals. As with direct investments, licensing proposals had to run the gamut of administrative regulations on TI.[8]

The regulations were designed to protect indigenous technologies wherever available; to keep down royalty rates (the government prescribed maximum rates, relaxed in very few cases, depending on the sophistication of the technology, of 3-5 percent of associated sales, and imposed a 40 percent tax on top); to reduce the life of the agreement (for a long time, the limit has been set at five years from the start of production, during which the Indian party was supposed to have undertaken R & D or similar measures to absorb the technology fully – renewals of agreements were cumbersome and difficult); and to permit the licensee to sublicense the technology locally. More broadly, the government prescribed areas where no foreign technology would be permitted and, within permitted areas, the particular products or processes that were desirable.

This complex and rigid structure of controls, often administered by officials who had little understanding of the technologies involved and even less of the dynamics of competition and efficiency, was set in an industrial regulation regime that sought to hold back the growth of large firms, protected existing producers regardless of efficiency, and completely protected the market from external competition for all time to come.

The results of this set of policies on TI via licensing were as follows:

First, a large number of approved agreements were never taken up. A study of a sample of 1,815 approvals in 1975-81 shows that only 37 percent of these were actually implemented.[9] If this could be generalized to the universe of agreements, only some 2,600 agreements (including those with equity investment) materialized in 1957-82, and about 1,200 purely technical ones in 1969-82. It is not clear why the rate of "failure" is so high, especially when obtaining approval is itself a difficult and lengthy process. Perhaps the stifling industrial regime within the country, the slow and erratic rate of growth, and the lack of real competitive pressure to improve productivity and introduce new products all contributed.

Second, the low royalty rates received by the licenser after the mid-1960s (1.8-3.0 percent after tax) undoubtedly meant that the quality of technology supplied suffered.[10] Leading innovators in the developed world are, in any case, reluctant to sell their most recent and profitable technologies to unrelated firms. They may be persuaded to do so if the royalty rates are high and the transfer is hemmed in by provisions to protect the licenser's main markets and long-term profit strategy. The Indian combination of low net royalties and requirements to permit sublicensing had the effect of inducing second- or third-grade technology to be sold in many instances. The ceiling on royalties was to some extent circumvented by including higher lump-sum fees, but this was also subject to controls and did not enable Indian firms to shop freely for the best technologies.

Tight controls on capital goods imports and rigorous indigenization requirements reduced the scope for paying for the technology in terms of products imported from the licenser.

Third, the rather static view of technology transfer inherent in the short life permitted to licenses (and the difficulties of renewing them) meant that licensees were unable to keep up with changes in technology. While many large firms did set up R & D units to assimilate and adapt licensed technologies and, with growing capabilities, could undertake the basic design of many products and processes, they did not have the independent ability to keep up with world frontiers in fast-moving technologies. In some cases, it is doubtful whether the five-year period was even sufficient to absorb a given complex technology fully. The interplay between domestic technological effort and technology import (the "make/buy decision" in technology development) is, of course, the crucial and little-understood issue underlying all these questions, and we shall return to it later.

As was noted earlier, the highly protected nature of the Indian market enabled producers to survive with obsolete or second-grade technologies in the areas of production allotted to them. At the same time, the pervasive control of the government over industrial investment and production meant that, even given these constraints, firms were unable to move into market-dictated areas of profitable activity. And, in addition to the disincentives to exporting inherent in the trade regime,[11] opportunities to exploit industry's emerging comparative advantage in international markets were not exploited.

How does India's import of licensed technology compare with that of other NICs? No centralized data on recent royalties and technical fees paid abroad are available for India. I have collected data on such remissions by the 433 largest private sector and 203 largest public sector companies from balance sheet data compiled by the Reserve Bank of India.[12] If one assumes that these firms account for most of the licensing agreements in the country, the most recent data on their technology payments are:

433 largest private companies (1979-80)	
Royalties	$ 11.0 m.
Technical fees	11.9 m.
	$ 22.9 m.

203 largest public sector companies (1978-79)	
Royalties	$ 57.8 m.
Technical fees	28.2 m.
	$ 85.0 m.
Total royalties	$ 68.8 m.
Total technical fees	40.1 m.
Total	$ 108.9 m.

We may infer that India's recent licensing payments are in the range of $100-120 million per annum. This compares with $782 million in 1979 for Brazil, $641 million in 1980 for Mexico, and $107 million in 1980 for Korea. If these payments are deflated by manufacturing value added (*WDR* figures for 1979), Mexico comes out as having the highest dependence on foreign licensed technology (2.7 percent), followed by Brazil (1.9 percent), South Korea (1.1 percent), and India (0.7-0.8 percent).

To complete the analysis of India's TI, it would be useful to touch on its purchases of embodied technology: capital goods. While I have deliberately excluded these from my definition of technology trade, my survey would be incomplete if they were entirely ignored.

In view of the quarter-century of efforts to achieve self-reliance in the production of capital goods, it is hardly surprising that India has the lowest value (absolute and relative to domestic production) of capital goods imports among the NICs. According to the Centre on Transnational Corporations' 1982 study, imports of engineering goods (this is a broad category including other products besides capital goods) comprised 20 percent of domestic production in 1970-71, and 15 percent in 1977-78. Its imports of capital goods (this is defined to exclude transport goods) came to $1.6 billion in 1978-79 compared to $4.4 billion for Brazil (1980), $3.5 billion for Mexico (1980), and $5.3 billion (1979) for Korea.[13]

Not only are direct imports of finished capital goods relatively low for India, but the import content of locally produced capital goods is also very small. According to Centre on Transnational Corporations estimates, India's average import content in this sector is under 10 percent compared to around 20 percent for Brazil, 45 percent for South Korea, and 50 percent for Mexico.[14] These figures conceal a great deal of variation around the mean, of course, but empirical observation confirms that India is able to undertake the full domestic manufacture of nearly all the critical, sophisticated components of industrial equipment and needs to import only special qualities of steel and some controls and special items that are not yet made locally. Brazil is very similar but has not pushed indigenization to quite the same lengths as India in advanced items. South Korea and Mexico, by contrast, tend to rely on imports of the more sophisticated components, especially for newer ranges of equipment.

To conclude, therefore, India has clearly managed to travel a long way in its quest for self-reliance in production and technology. The journey has not been costless. Large areas of industry are uncompetitive because plants operate with varying mixtures of outdated technologies, uneconomic scales, high levels of x-inefficiency, and poor quality control. They are, in addition, hampered by high-cost inputs, gross infrastructural deficiencies, poor labor relations, and a rigid, cumbersome, often corrupt regulatory superstructure. This is the negative side. On the positive side, there *are* individual firms in the engineering sector that are efficient by world standards, that can competitively provide some technologies that are in international demand, and that judiciously combine TI with their own R & D to keep up with changing technologies. More generally, there has been a broad-based effort to assimilate and adapt technologies. India still enjoys the largest and cheapest pool of engineering and scientific manpower in the Third World (in 1977-78 there were 1.8 million scientists and engineers employed there plus another 0.2 million unemployed, compared with stocks of about 0.6 million in Brazil, 0.8 million in South Korea, and 0.4 million each in Mexico and Argentina),[15] one of the largest in the world. This provides a large absorptive base for technology as well as a very competitive resource for deployment abroad. The peculiar nature of technological development in India and its human endowments have resulted in considerable TE, to which I now turn.

TECHNOLOGY EXPORTS BY INDIA

India is not a major capital goods exporter among the NICs. In 1978, the total value of product exports in this category came to $421 million, compared to $1.4 billion for Brazil, $1.5 billion for South Korea, and presumably a similarly high figure for Taiwan. Argentina and Mexico came slightly behind India with $407 million and $364 million respectively. If, however, one considers exports of technology per se, or accompanied by the sale of capital goods, India emerges as one of the leaders, in terms of value of exports and/or the range, depth, and complexity of technologies provided.

As far as the setting up of industrial projects overseas is concerned, my collection of scattered pieces of data shows that India has won well over 200 contracts abroad (mainly as prime contractor to design, engineer, provide capital goods, erect, and commission industrial projects, including power generation and transmission). The total value of contracts won is not known but lies between $2 and $2.5 billion, mainly for the period 1975 to early 1982. This may be compared to similar project contracts won by Brazil of $111 million (1966-80), Argentina $160 million (1973-80), and Mexico $43 million (1974-79).[16] South Korean data are difficult to interpret because they are not collected on a comparable basis (they include "plant exports" where no engineering/technical services are undertaken overseas). Leaving out exports of plant (that is, equipment) to developed countries, offshore oil rigs, and coastal facilities, none of which are TE in the sense used here, we are left with a total of $802 million (cumulative to mid-1981) for plant exports.[17] Even this is an obvious overestimate, because *all* plant exports to the Third World are counted as projects executed abroad by South Korea, and this may often not be true. In particular, the export of power generation plant (worth $110 million) is very likely to have taken the form of straightforward product exports, and items like "onshore structures" (worth $98 million to developing countries) are not really industrial exports. But even if one gives or takes a few items like this, South Korean turnkey industrial exports are less than half of India's. (This is in contrast to their civil construction exports which are seven to eight times larger.)

In the area of technology exports by means of direct investment,[18] Hong Kong leads the Third World with indigenous (as opposed to expatriate British firms or *hongs*) holding a capital stock of some $600-700 million overseas. Most of this appears to be in manufacturing industry, aimed at setting up export-oriented plants in other Asian countries in the more mature of Hong Kong's export products – textiles, simple garments, plastics, consumer electronics,

TABLE 3.1
India: Distribution by Industry of Industrial Projects, Completed and in Hand, Mid-1982
(Rupees million)

	Projects with Data on Values						*Percent Distribution of Value*	*Projects without Data*		
	Total		*Public Sector*		*Private Sector*					
Industry	*Number*	*Value*	*Number*	*Value*	*Number*	*Value*	*(Total)*	*Total*	*Public Sector*	*Private Sector*
Textile, yarn[1]	20	1,657.25	3	339.30	17	1,317.95	9.9	4	1	3
Sugar	12	851.43	3	365.56	9	485.87	5.1	3	–	3
Other food processing	4	36.10	1	4.20	3	31.90	0.2	2	1	1
Cement	6	1,909.10	2	1,080.00	4	829.10	11.4	2	1	1
Chemicals	4	56.20	2	45.30	2	10.90	0.3	Several	1	Several
Paper & pulp	2	92.51	1	49.41	1	43.10	0.6	1	–	1
Simple metal products	8	511.93	1	14.00	7	497.93	3.1	1	–	1
Steel mills, other metals	11	1,595.06	7	489.86	4	1,105.20	9.5	2	–	2
Machinery & machine tools	10	1,219.80	10	1,219.80	–	–	7.3	4	3	1
Power generation[2]	53	3,876.72	16	3,023.70	37	853.02[3]	23.2	10	4	6
Power distribution	n.a.	3,154.82	2	31.90	n.a.	3,122.92[4]	18.9	2	1	1
Transport equipment	1	70.00	–	–	1	70.00[5]	0.4	Several	1[6]	Several
Electronics[7] telecommunication	8	227.20	7	225.70	1	1.5	1.4	5	5	–
Miscellaneous	23	57.50	21	21.7[8]	2	35.8[9]	0.3	Several	–	Several
SUBTOTAL	>162	15,315.62	76	6,910.43	>88	8,405.19	91.6	>36	18	>20

Steel structures, boilers & tanks	24	895.19	6	383.50	18	511.69	5.4	Several	3	Several
Water treatment, sewage plants	17	477.86	4	237.00	13	240.86	2.9	1	–	1
Small-scale units	n.a.	30.00	n.a.	n.a.	n.a.	n.a.	0.1	Several	–	–
TOTAL	>203	16,718.67	86	7,530.93	>119	9,157.74	100.0	>37	21	21
% Distribution	–	100.0	–	45.0	–	54.8	–			

Notes:

[1] Includes viscose fibre plants (Rs.563m) undertaken by Gwalior Rayon in S. Korea and the Philippines.

[2] Includes substations, transformer stations and power control panels.

[3] Includes 9 contracts (Rs525.9m) for airconditioning & refrigeration equipment.

[4] Number of contracts is not available, but Kamani Engineering Corporation accounts for Rs.1777.4m and Tata Exports for Rs.542.lm. of the total.

[5] Bicycle plants, commercial vehicle plant.

[6] Scooter plants.

[7] Telephone exchange and ancillaries.

[8] 21 pilot plants including electronics, metallurgy, food, paper, leather, phytochemicals and plastics, undertaken by FIDMA.

[9] Conveyor, Ash Dust handling system.

Sources: Various reports, newspapers, balance sheets, interviews.

and the like. These are the activities that are facing the most intense competition from the lesser-industrialized countries of Asia, where rising land and labor costs encourage relocation in lower-cost countries. The more design- or technology-based products (such as fashion garments, toys, and sophisticated electronics) can be economically retained at home.

Hong Kong is unique among the NICs in undertaking primarily this "dynamic comparative advantage" type of foreign investment. Other investors are far more geared to promoting exports from the home base and meeting the import-substitution needs of the host country. Even the investments of Singapore, which would ostensibly be subject to similar push factors, tend to be directed at meeting ethnic and host country needs. Singapore's own exports are, of course, much more in heavy, high-technology products than Hong Kong's and are dominated by developed country MNCs to a much greater extent (over 90 percent of its exports arise from MNCs as compared to around 15 percent for Hong Kong). Singapore is still a substantial investor in neighboring countries like Malaysia and Indonesia.

Neither of these island economies exports much by way of industrial technology (in the form, that is, of locally made capital goods or basic production engineering and design). What they provide is mainly entrepreneurship, production know-how, and (especially for Hong Kong firms) aggressive marketing skills and contacts in international markets: the ability, in other words, to commercialize a given technology efficiently rather than to reproduce the technology itself. Given the absence of a large indigenous capital goods sector, this is hardly surprising. And given the competitive acumen of their enterprises, the skills and know-how they do possess more than compensate for their lack of basic technologies.

Among the larger Third World economies, foreign investors tend to base themselves more on their domestic technology, in terms of capital goods, basic product/process design, and supporting engineering services. In the manufacturing sector, India emerges as the largest investor overseas, with a total equity stock of over $100 million and a manufacturing equity stock of approximately $85 million. If one compares this to earlier figures on direct investment inflows in the 1970s, the odd result is that one of the world's poorest countries has become a net exporter of direct investment. Whether this is a fact to be proud of – the government has succeeded in keeping out foreign MNCs and building up national enterprises – or a sad reflection on the state of the Indian economy is another matter.

Of the other NICs, South Korea, Taiwan, Argentina, and Mexico have foreign equity in manufacturing of around $30-60 million each.* Brazil has even less, around $20 million, though its nonmanufacturing investments overseas (in the petroleum and civil construction sectors) are enormous, in the region of $1 billion.

India's relatively large foreign equity stake (the island economies aside) is not the only notable feature: the diversity and complexity of its foreign investments are also much greater than for the other NICs. Indian MNCs are engaged in everything from traditional, low-technology industries like textiles and food processing (though many of the facilities themselves have fairly modern equipment) to large-scale, sophisticated ones like truck assembly, jeep assembly, pulp and paper, rayon, carbon black, palm oil fractionation, minicomputers, and precision tools. Table 3.2 shows the broad industrial distribution of Indian foreign investments.

Indian regulations require that the capital contribution made by Indian investors generally take the form of local equipment and capitalized know-how. Cash contributions are rarely permitted in manufacturing ventures (this creates financial problems for some affiliates). Thus, while a part of the equipment used is bought from developed countries, the core technology is essentially contributed by India. Most Indian investments are directed to the home markets of the host countries, but there are several exceptions. The pulp and paper mill in Kenya, the largest in developing Africa, exports one-fourth to one-third of its output; the palm oil fractionation plants in Southeast Asia are primarily export oriented; most investments in Singapore (precision tools, steel tubes, minicomputers, and so on) are also mainly export oriented, as are some being set up in Sri Lanka's duty-free industrial zone.

---

*The Argentine data need to be treated with caution. They relate only to direct investments made in the past two decades; however, in the earlier part of this century three large Argentine firms went multinational in a big way. The affiliates later grew away from the parents, and two ended up being larger and more powerful. These peculiar cases are excluded from the calculations, but for longer analysis see Katz and Kosacoff, "Multinationals from Argentina," in S. Lall et al., *The New Multinationals: The Spread of Third World Enterprises* (Chichester: J. Wiley, 1983).

TABLE 3.2
Industrial Distribution of Indian Direct Foreign Investments in Operation and under Implementation, End August 1980

(rupees million)

	Ventures in Operation			*Ventures under Implementation*			*Total*		
Activity	*No.*	*Indian Equity (value)*	*%*	*No.*	*Indian Equity (value)*	*%*	*No.*	*Indian Equity (value)*	*%*
Manufacturing									
Textiles, yarn, etc.	16	101.6	28.6	10	133.8	23.5	26	235.4	25.4
Sugar	1	11.4	3.2	2	26.0	4.6	3	37.4	4.0
Food and palm oil processing	20	71.1	19.9	8	30.9	5.5	28	102.0	11.0
Iron & steel mills	5	11.7	3.3	1	2.2	0.4	6	13.9	1.5
Chemicals, drugs	9	16.3	4.6	8	12.1	2.1	17	28.4	3.1
Paper & pulp	2	52.3	14.6	3	114.1	20.0	5	166.4	18.0
Engineering	22	63.9	17.9	15	92.9	16.3	37	156.8	16.9
Leather, rubber	2	0.8	0.2	1	3.2	0.5	3	4.0	0.4
Glass	3	6.1	1.7	1	6.8	1.2	4	12.9	1.4
Subtotal	80	335.2	93.8	49	422.0	74.1	129	757.2	81.7

Non-Manufacturing									
Hotels & restaurants	14	3.7	1.0	9	68.3	12.0	23	71.9	7.8
Consultancy	6	0.6	0.2	4	3.0	0.6	10	3.6	0.4
Eng. & construction contracting	4	2.9	0.8	9	33.8	5.9	13	36.6	4.0
Trading	9	4.8	1.3	7	1.5	0.3	16	6.3	0.7
Shipping	1	2.2	0.6	1	0.8	0.1	2	3.0	0.3
Mineral exploration	—	—	—	2	22.0	3.9	2	22.0	2.4
Miscellaneous	3	7.9	2.2	6	18.1	3.2	9	26.0	2.8
Subtotal	37	21.9	6.2	38	147.4	25.9	75	169.4	18.3
Total	117	357.1	100.0	87	569.4	100.0	204	926.5	100.0

Source: Indian Investment Centre, *Indian Joint Ventures Abroad: An Appraisal*, New Delhi: IIC, 1981.

The spread and complexity of Indian MNCs raise interesting issues for the current theories on Third World multinationals. Certainly, Indian firms are something of an exception to the generalization that such MNCs tend to be small, labor intensive and low technology in their industrial activities.[19] They do lack marketing know-how (in terms of having highly advertised, differentiated goods), and tend to stay mainly in intermediate and capital goods manufacture, but in terms of the technological requirements of their activities they span a wide range.

Let us now come to consultancy. India has approximately 60 professional consultancy organizations that have exported their services, and about half of these are regularly engaged in overseas projects. The accumulated value of their earnings (until April 1982) is estimated to be about $125 million. If one extrapolates from the data for 1978-79, about half of overseas earnings are in the manufacturing sector (if one counts power generation and computer software as nonmanufacturing). Thus, just over $60 million was earned from the export of industrial consultancy mainly in the form of feasibility studies, project monitoring, and detailed engineering, that is, the "pure consultancy" required for setting up industrial projects. The main sectors for manufacturing consultancy were iron and steel and chemicals, with a handful of large public and private sector firms gaining a fairly secure foothold in developing country markets in Africa, Asia, and Latin America.

It is extremely difficult to get comparable data on the earnings of professional consultancy firms from the other NICs. According to Inter-American Development Bank data for 1982, consultancy earnings – manufacturing *and others* – by the three large Latin American countries were $10 million for Argentina (1973-77), $8.4 million for Brazil (1966-80), and $9.4 million (1974-80) for Mexico, but these are estimates based on partial data.[20] Nevertheless, the broad magnitudes indicated suggest that Latin American consultants are much less active overseas than their Indian counterparts.

South Korean data would seem to suggest the opposite for South Korean consultants. According to a survey done by Amsden and Kim for the World Bank technology exports project, the accumulated value of "technical consultancy" exports by South Korea (until the end of 1981) came to $306 million, of which $155 million arose in the manufacturing sector. A closer look reveals, however, that the data are not really comparable with those for India. First, the

South Korean data include such technical services as operation and maintenance (in 1980 these accounted for about 82 percent of the exports), which are not covered by the Indian consultancy data. Second, the data seem to be based upon the value of contracts registered with the Ministry of Science and Technology, rather than actual remittances in the relevant period, as is the case with India. Both tend to overstate greatly the relative value of South Korean consultancy exports. Thus, if 20 percent of the value of the manufacturing consultancy went into feasibility studies, design, and similar "pure consultancy" work (using the 1980 breakdown as a guide) we get a total of $31 million; if about half of this is actual earnings, we get a rough total of $16 million; this may be the more appropriate figure to compare to India's $60 million.

When we come to the final category of TE, licensing and technical services, we find a complete absence of centralized data on India: a strange lacuna in a country that otherwise collects immense amounts of data on all aspects of industrial activity. The best we can do is to take the Reserve Bank of India's (RBI) compilation of balance sheet data for the largest public limited companies on actual earnings of royalties and technical fees overseas. This tends to understate the actual earnings insofar as other firms are left out (many medium-size firms undertake technical assistance work overseas) and the total values of the contracts are not shown. It tends to overstate them to the extent that some earnings double-count the value of services already included in project exports. Still, some data are better than none. The 300-400 largest private sector firms earned $2.5 million in royalties and $62.3 million in technical fees in the period 1975-76 to 1980-81. Public sector enterprises earned some $58 million in technical fees in 1975-76 to 1978-79, but a large element here seems to be part of turnkey jobs abroad.

Since these are again earnings rather than contract data, they do not reflect some substantial licensing contracts (for example, $6.5 million for Hindustan Machine Tools, $6.1 million for Bharat Heavy Plates and Vessels, $4.3 million for Heavy Engineering Corporation, and $2.5 million for Scooters India Limited in the public sector, and $27.5 million for Associated Cement Companies, $50 million for the Indian Aluminum Company, $33.8 million for the Electrical Construction Company in the private sector) by Indian standards.[21] Because of the large gaps in the data it is impossible to arrive at a total estimate.

Nor do other NICs provide data on licensing earnings overseas. The only exception is South Korea, which, as noted earlier, enables data on total contract values to be calculated. Amsden and Kim arrive at a total of $139 million worth of technical licensing (royalties plus lump sum payments) for the manufacturing sector until the end of 1981. To this we should add the $124 million worth of maintenance and operation contracts shown under technical consultancy getting a total of *$263 million* for what I have defined as licensing. This may be larger or smaller than total accumulated contract values for similar technology sales by India; we simply cannot tell.

Let us sum up this section. TE by India had grown to impressive proportions in the late 1970s, not just in absolute values but also in the diversity and complexity of technologies they embody. While other NICs may sometimes have larger amounts of particular forms of TE (for example, Hong Kong in direct investment and perhaps South Korea in technical services), none of them matches India in terms of overall performance, depth, and range of indigenous technological capability in manufacturing industry. India is able to provide not just the operating knowledge to set up and run industries (the know-how), but also the design and manufacture of the plant and equipment, designed specifically for the client (the "know-why"): it is in this latter sense that its TE seem particularly noteworthy in the context of the NICs.

But let us not forget the larger setting. As I have pointed out already, India has a much worse export performance than many other NICs, in general and in the area of capital goods. Its growth record has been equally poor. There are obvious areas of inefficiency and technological backwardness in its industry. And, most interestingly, it has imported the least new technology in recent periods. Its TE performance per se would simply point to growing industrial maturity and a cost advantage in technical manpower. It is the configuration of such maturity with various indexes of inefficiency and technological lags that raises the most difficult and intriguing questions. Let us turn to these in the next section.

AN ASSESSMENT OF INDIA'S PERFORMANCE

In a preliminary analysis of India's TE I had argued that its relatively high TE and relatively low TI were causally connected.[22]

India's low reliance on foreign technologies, together with its effort to boost indigenous R & D and local enterprise, represented a policy of "protection of technological learning" that was quite distinct from its protection of local production. If TE could be used as an index of indigenous technological capability, India had developed arguably the most diverse and deep (in terms of going into basic design of products and processes) capabilities among the NICs.

I took it for granted that this buildup of capabilities was a good thing, for several reasons (for example, more appropriate technology, cheaper technology, a better bargaining position for buying new foreign technologies, dynamic and external effects, realization of dynamic comparative advantage, and so on). With the recent historical example of Japan – the most successful practitioner of the policy of protecting technological learning – exercising a very strong influence on my reasoning, it seemed obvious that a deep technological base could provide the basis for sustained long-term growth. I saw clearly the inefficiencies in the Indian import-substitution and other regulatory policies that had accompanied the buildup of technological capabilities. I suggested, nevertheless, that these inefficiencies were not a necessary part of the policies for the protection of technological learning. A different set of industrialization policies (à la Japan), combined with selected policies to promote indigenous technology, could have kept the benefits of the latter while avoiding the excesses of Indian regulations.

After the benefit of four years of field work on Indian industry, I feel compelled to qualify this argument. There is still no reason to doubt that India *has* built up a great deal of technological capability in the past three decades of protected industrialization and restricted technology inflows. There are some firms that are competitive exporters of technology in their own right (that is, without much government support). In certain branches of industry (cement and paper plant up to certain capacities, medium-size power stations, some textiles and some simpler engineering products),[23] India can set up good industrial plants abroad and provide relevant technical service. Effects of this technological development have probably spread through vertical linkages. India now has a much better absorptive base for injections of fresh technologies from abroad.

Despite these benefits, the set of policies that has supported India's technological development has been particularly costly and has not endowed it with a base for dynamic growth in the future.

To support this interpretation, it may be instructive to retrace two main elements in my earlier argument.[24]

First, the need for and desirability of the protection of learning. A certain amount of know-how accumulation is a necessary part of the production process, because of the need to introduce new techniques into production, iron out difficulties, make the inevitable adaptations to local conditions, train workers, and the like. No particular strategy is required to foster this – local or foreign enterprises, domestic or export-oriented firms, all generate such know-how. The progression to deeper technological capability – design changes, basic design capability, new product/process development – requires investments in accumulating know-why. In developing countries, even the accumulation of know-why for technologies well diffused in the industrial world is a costly and risky process. Faced with the alternative of easily available (licensed or transferred by direct investment), proven foreign technologies, most firms would probably not undertake the risk and cost involved. To a certain extent, therefore, indigenous effort and TI are substitutes. Insofar as domestic know-why can be competitively developed, government intervention to reduce access to predigested foreign technologies may be recommended to bring private and social benefits of technological effort into line.

This is the case for protecting know-why development. But there are two snags. First, since technical resources are limited (even in countries like India, for manpower of high quality), a diversion of these resources to know-why activities may reduce the manpower available to assimilate production know-how. Second, it may be counterproductive to spend a lot of effort in reaching a given level of know-why when the technology is subject to rapid change internationally. The developing country firm may, after years of design work, high costs, and mistakes, find that its ability to reproduce a given technology from first principles has become obsolete. It may be able to sell this capability to lesser-industrialized countries, but neither the know-why nor its output are viable in developed country markets, and neither can compete against more recent technologies deployed by other NICs.

The crucial point to note here is that, in developing country enterprises, the mastery of a certain level of know-why may not imply the capability to develop it further in line with developments abroad. In sectors where competitiveness requires that NICs deploy increasingly sophisticated techniques (and this is the case in most manufacturing industry), a strategy of technological self-reliance of the Indian type may stultify further technological development. My field work in India has convinced me of the very real limitations to know-why development in even the largest enterprises there, and I believe that these apply to most Third World enterprises. There is, in other words, a very real scale of comparative advantage in the development of technological capability. Once that advantage is exhausted, it is undesirable to push learning further.

India has clearly pushed it much further, condemning large sectors of industry to technical obsolescence. It is not that the enterprises concerned have not tried to develop independently (though many have not), but that they have reached the limits to their learning. Nor is it to argue that *any* know-why development is inadvisable in developing countries, but only that it has to be continuously sustained by injections of fresh foreign technologies. This, indeed, is the lesson of Japan's technological strategy. As each new stage of know-why development is reached, the country's potential for independent growth is larger, but true international viability cannot be reached until a much later stage of development. In India's case, its best firms find markets for their accumulated technology in some Third World countries, but these are small (and shrinking) markets: technology exports and technological development can, in other words, coexist with growing technological backwardness.

The relationship between TI and local technological effort is therefore a continuously varying one.[25] At certain stages the two are substitutes, and intervention is required to bring private efforts into line with social needs. At others, they are complementary. As the economy develops, the need for intervention is correspondingly reduced. Not only do indigenous enterprises see the need for greater technological effort to assimilate imported technologies[26] and to develop new ones to suit peculiar local circumstances; but foreign enterprises also respond to larger, more sophisticated (and better skill-endowed) local markets by investing in local know-why.[27] Some tension betwen the two always remains – witness the technology policies of Western European countries today – but the need for intervention is much more specific and (hopefully) shortlived.

The optimum trade-off between imported and domestic technologies will also vary greatly from one industry to another. In relatively easy or stable technologies, local know-why can be carried much further without loss of efficiency than in complex, large-scale or rapidly changing ones. India's self-reliance efforts here, unfortunately, pushed out indiscriminately. The result has been that it has built up genuine advantages in a few easy sectors and lost out in several difficult ones.

The first major qualification is, then, that a strategy of low TI may lead to a certain buildup of capabilities (which spill over into TE). When carried too far, however, such a strategy may lead to technological stultification because of inherent limits on what developing country enterprises can do on their own. Their capabilities can grow, no doubt, but so slowly (and at so great a cost) that society loses in relation to its more liberal competitors. The export of some of these capabilities to lesser-industrialized countries should not conceal the fact that many of them are too obsolete to be beneficial to the exporting country itself.

The second main qualification concerns the relationship between technological and other interventionist policies in India. We can treat the two separately only if the nature of technological change is independent of the nature of other distortions in the economy (that is, if all technological development is equally good for the economy) and if the pace of technological development is also similarly independent (that is, firms push themselves to their technological limit regardless of the trade regime and other controls). Neither assumption may be valid.

The nature of technological effort in India has been strongly conditioned by the nature of the government's policy interventions. While some of the past efforts were necessitated by local circumstances and market needs outside the government's control, many others were undertaken specifically because of policy interventions: to indigenize inputs rapidly, to subcontract to small-scale suppliers, to cope with infrastructural bottlenecks, and to utilize second-rate imported technologies. Much of the latter category of effort may have been socially wasteful.[28] In a more liberal, efficient, and outward-looking regime it might not have been necessary. As development economists we may be too prone to applaud *any* increase in technological capability; perhaps we should acknowledge that some species may be undesirable.

The pace of technological change can also be influenced by the trade regime. The literature on innovation in industrialized countries[29] shows that the strength of competition is among the crucial factors determining the propensity to undertake technological change, by innovation, imitation, or licensing. By severely constricting competition from the most modern technologies (as represented by imports or MNCs), India has encouraged or permitted its firms to be technologically slothful. Sloth has several aspects here: first, little independent effort to go beyond the minimum effort to put into operation imported technologies in Indian conditions; second, no effort to import new generations of technologies, even when more liberal policies (since 1980) on TI have been introduced; third, the making do with second-grade imported technologies because these were the only ones that Indian regulations would permit. This is not to deny that some exceptional firms have been technologically dynamic, but these are overwhelmed by the mass of other firms that have not. For these others, it is not just the supply of foreign technology that has been constricted. Over time their demand for new technology imports has also fallen. This is a sign not of growing competence but of profitably sheltered inefficiency.

Furthermore, it is likely that a part of the technological effort that has taken place in India has been dissipated because of other regulations. The constraints placed on the growth of large firms within particular markets, the encouragement of small-scale producers, the emphasis on public sector investment despite its abysmal record in several industries, the infrastructural constraints on production (due to poor government planning and performance), must all have frustrated the more innovative firms in the exploitation of their efforts.

Another consequence of this set of regulations on large private firms has been that some firms have found it preferable to export technology from India (and set up operations overseas) than to export their products. This applies more to foreign direct investment than to other forms of TE, and its importance should not be overemphasized: most Indian MNCs still have marginal proportions of their total assets overseas. Nevertheless, the motivation to escape the tightly controlled home economy is present and must be mentioned. Add to this the costs, delays, and difficulties in exporting products from India, and it becomes understandable why a number of firms find the TE route easier.

It is now possible to detect some of the complex links between low TI, high TE, and poor industrial and export performance on the part of India. The evidence is still scanty and highly impressionistic, and my interpretation may be contentious. Nevertheless, to sum up, this chapter has suggested that technological protection and import substitution *have* generated a great deal of indigenous technological effort in India. This effort has enabled many enterprises to build up basic design and project execution capabilities, which have resulted in substantial TE. At the same time, the growth of technological capabilities has been limited by some inherent constraints on Third World enterprises, which have forced many technologies to become progressively obsolete. These technologies find markets in lesser-industrialized countries, but India itself would benefit by moving to the more advanced technologies being deployed by other NICs. Apart from building up basic know-why, however, the Indian policy regime has also generated a lot of wasteful technological effort and has dampened the competitive need to import, implement, and build upon more efficient technologies. The simultaneous existence of technological capability and backwardness, of TE and uncompetitiveness in product exports, of technological effort and sloth, is part of the series of paradoxes that characterize India. Given the base of human resources and the mélange of contradictory stimulating and inhibiting policies, perhaps this was only to be expected.

CONCLUSION

This chapter has attempted to go beyond a simple analysis of technology trade by India. It has attempted, admittedly on the basis of scanty evidence, to link together the extent of technology imports and exports with the growth of indigenous capabilities, and to set the whole complex process in the broader context of sluggish industrial performance and widespread intervention. A short chapter like this cannot be expected to do justice to this set of issues. I am conscious of many questions left untouched and the cursory treatment of many complex phenomena.

I will not attempt to summarize the discussion of India's technology policies and technological development. Let me only highlight some points that may be of more general interest.

First, a low reliance on imports of technology in the process of industrialization clearly contributes to the buildup of a diverse and deep technological capability. Without going into the costs and efficiency of such a strategy, I must admit that India's technology exports suggest that, despite its other handicaps in exporting and exploiting its comparative advantage in product trade, it has built up an impressive stock of know-why for a poor country in comparison to countries that have imported technology more freely.

Second, India seems to have erred in its technological development in two ways: 1) Its emphasis on self-reliance in technology has forced enterprises to fall behind international frontiers in many areas. Third World firms can assimilate and adapt imported technologies, but they cannot keep up unaided with the pace of major innovation in the developed world. There is a scale of comparative advantage in technological development, and India has attempted to go beyond its capabilities. 2) Its highly protected domestic market, with disincentives to export activity and a rigid superstructure of inefficient controls, has also reduced the demand for new foreign technologies.

Third, technology exports to lesser-industrialized countries do show a certain level of technological competence, based on production experience, adaptation to local conditions, and design and development of know-why. In some cases, the technology supplied by India may be comparable to world frontiers (in stable-technology industries); in others, its selling point is its outdated, simplified, small-scale, or adapted nature. In the latter category, India's technological competence may be a positive handicap to raising productivity at home, because its user industries are stuck with technologies that do not allow them to compete with other NICs. India should, in other words, be moving to far more advanced technologies than its own capabilities can provide, and TE is a sign of its slow progress.

Fourth, a certain proportion of the technological competence engendered by the Indian policy regime has been beneficial, in the sense that it has provided a base on which new techniques can be grafted and developed. The rest of it has been socially wasteful, in that it has been generated in response to policy interventions that were neither necessary nor in the long-term interests of the economy. Not all technological development is desirable: only that which takes a country along a dynamic growth path in keeping with its changing comparative advantage.

Finally, while this chapter has not gone into policy recommendations, it is clear that the analysis points toward much greater liberalization of policy in India – not just in technology imports, since injections of new technologies in the present protected, inward-looking, fragmented structure of industry would not really be desirable, but liberalization in the industrial policies that have fostered this structure. I am not optimistic about this outcome. Despite a great deal of talk in India about sweeping policy reforms, the changes are only cosmetic. There are too many vested interests in the existing structure of private industry and the government to permit any substantial change without a much greater upheaval than is now evident. The final lesson – it is so well known as to be almost banal – is precisely this: a misdirected strategy creates the sociopolitical structure for its own preservation, despite its long-term costs to society.

NOTES

1. S. Lall, *Developing Countries in the International Economy* (London: Macmillan, 1981); F. Stewart, "International Technology Transfer: Issues and Policy Options," World Bank Staff Working Paper no. 344 (1979).

2. S. Lall, *Developing Countries as Exporters of Technology: A First Look at the Indian Experience* (London: Macmillan, 1982).

3. N. Rosenberg, *Perspectives on Technology* (Cambridge: Cambridge University Press, 1976).

4. J. Katz, "Technological Change and Development in Latin America," in *Latin America and the New International Economic Order*, ed. R. Ffrench-Davis and E. Tironi (London: Macmillan, 1982); C. Cooper, "Policy Interventions for Technological Innovation in Developing Countries," World Bank Staff Working Paper no. 441 (1980); C. Dahlman and L. E. Westphal, "Technological Effort and Industrial Development – an Interpretative Survey of Recent Research" in *The Economics of New Technology in Developing Countries*, ed. Frances Stewart and Jeffrey James (London: Frances Pinter, 1982); M. Fransman and K. King, eds., *Technological Capability in the Third World* (London: Macmillan, 1984); Lall, *Developing Countries as Exporters of Technology*; S. Lall and S. Mohammad, "Technological Effort and Disembodied Technology Exports: An Econometric Analysis of Inter-Industry Variations in India," *World Development* 11 (1983): 527-35; L. E. Westphal, "Empirical Justification for Infant Industry Protection," World Bank Staff Working Paper, no. 445 (1981); L. E. Westphal and Y. Rhee, "Korea's Revealed Comparative Advantage in Exports of Technology: An Initial Assessment," (World Bank, 1982) mimeograph; S. Teitel, "Towards an Understanding of Technical Change in Semi-Industrialized Countries," *Research Policy* 10(1980): 127-47.

5. This section draws heavily on Indian Investment Centre, "Changing Forms of Foreign Involvement in India" (New Delhi: IIC, 1982) mimeograph; A. V. Desai, "Technology Import Policy in the Sixties and Seventies: Causes

and Consequences" (New Delhi: National Council of Applied Economic Research, 1982), mimeograph.

6. J. N. Bhagwati and P. Desai, *India: Planning for Industrialisation* (London: Oxford University Press, 1970); J. N. Bhagwati and T. N. Srinivasan, *Foreign Trade Regimes and Economic Development: India* (New York: National Bureau of Economic Research, 1975).

7. United Nations, Centre on Transnational Corporations, "Technology Transfer through Transnational Corporations in Capital Goods Manufacture in Selected Developing Countries" (New York: United Nations, Centre on Transnational Corporations, 1982), 2: 160.

8. S. Lall, "India's Technological Capacity: Effects of Trade, Industrial and Science and Technology Policies," in Fransman and King, *Technological Capability*.

9. Indian Investment Centre, "Changing Forms of Foreign Involvement in India" (New Delhi: Indian Investment Centre, 1982), p. 24.

10. Desai, "Technology Import Policy"; U.N., Centre on Transnational Corporations, "Technology Transfer through Transnational Corporations in Capital Goods Manufacture in Selected Developing Countries" (New York: U.N., Centre on Transnational Corporations, 1982), mimeograph.

11. Bhagwati and Srinivasan, *Foreign Trade Regimes*; B. Balassa et al., *Development Strategies in Semi-Industrial Economies* (Baltimore: Johns Hopkins Univesity Press, 1982).

12. Dahlman and Westphal, "Technological Effort and Industrial Development."

13. U.N., Centre on Transnational Corporations, "Technology Transfer through Transnational Corporations."

14. Government of India, *Draft Five Year Plan 1978-83* (New Delhi, Government of India, 1978); Dahlman and Westphal, "Technological Effort and Industrial Development."

15. Ibid.; Lall, *Developing Countries in the International Economy*, chap. 7.

16. Inter-American Development Bank, "Technology Exports from Latin America," in *Economic and Social Progress in Latin America: The External Sector* (Washington, D.C.: IDB, 1982).

17. Westphal and Rhee, "Korea's Revealed Comparative Advantage."

18. S. Lall, "Multinationals and Technology Development in Host LDCs," paper presented to the seventh world congress of the International Economics Association, Madrid, September 1983, on Structural Change, Economic Interdependence and World Development (transactions forthcoming).

19. S. Lall et al., *The New Multinationals: The Spread of Third World Enterprises*.

20. Inter-American Development Bank, "Technology Exports from Latin America"; Dahlman and Westphal, "Technological Effort and Industrial Development"; Dahlman's survey of Mexico estimates that total consultancy and licensing fees earned by Mexico came to $51 million (until October 1981), but this does not separate these quite distinct forms of TE, which we treat differently here.

21. An illustrative list of licensing contracts by India, but with large numbers of missing values, is shown in S. Lall, "India," *World Development*, special issue on 'Exports of Technology by Newly Industrializing Countries,' 12 (1984): 535-65.

22. Lall, *Developing Countries as Exporters of Technology*.

23. Cooper, "Policy Interventions for Technological Innovation"; these are stable mechanical engineering technologies where minor innovation is relatively easier than in chemical or electronics technologies.

24. Lall, *Developing Countries as Exporters of Technology*.

25. T. Blumenthal, "A Note on the Relationship between Domestic Research and Development and Imports of Technology," *Economic Development and Cultural Change* 27 (1979): 303-6, for a more general argument about the complementarity of imported and indigenous technologies.

26. D. J. Teece, *The Multinational Corporation and the Resource Cost of Technology Transfer* (Cambridge, Mass.: Ballinger, 1976), for a description of how local R & D greatly reduces the costs of technology transfer.

27. Lall, "Multinationals and Technology Development in Host LDCs."

28. See Teitel, "Towards an Understanding of Technical Change in Semi-Industrialized Countries," for a consideration of intervention-induced wasteful technological effort in the Latin American context.

29. M. I. Kamien and N. L. Schwartz, *Market Structure and Innovation* (Cambridge: Cambridge University Press, 1982).

REFERENCES

B. Balassa and Associates. *Development Strategies in Semi-Industrial Economies*. Baltimore: Johns Hopkins University Press, 1982.

J. N. Bhagwati and P. Desai. *India: Planning for Industrialisation*. London: Oxford University Press, 1970.

J. N. Bhagwati and T. N. Srinivasan. *Foreign Trade Regimes and Economic Development: India*. New York: National Bureau of Economic Research, 1975.

T. Blumenthal. "An Note on the Relationship between Domestic Research and Development and Imports of Technology." *Economic Development and Cultural Change* 27 (1979): 303-6.

C. Cooper. "Policy Interventions for Technological Innovation in Developing Countries." World Bank Staff Working Paper no. 441, 1980.

C. Dahlman and L. E. Westphal. "Technological Effort and Industrial Development – An Interpretative Survey of Recent Research." In *The Economics of New Technology in Developing Countries*, edited by Frances Stewart and Jeffrey James. London: Frances Pinter, 1982.

A. V. Desai. "New Forms of International Investment in India." New Delhi: National Council of Applied Economic Research, 1983. Mimeograph.

———. "Technology Import Policy in the Sixties and Seventies: Causes and Consequences." New Delhi: National Council of Applied Economic Research, 1982. Mimeograph.

M. Fransman and K. King, eds. *Technological Capability in the Third World.* London: Macmillan, 1984.

Government of India. *Draft Five Year Plan 1978-83.* New Delhi: Government of India, 1978.

Indian Investment Centre. "Changing Forms of Foreign Involvement in India." New Delhi: IIC, 1982. Mimeograph.

Inter-American Development Bank. "Technology Exports from Latin America." In *Economic and Social Progress in Latin America: The External Sector.* Washington, D.C.: IDB, 1982.

M. I. Kamien and N. L. Schwartz. *Market Structure and Innovation.* Cambridge: Cambridge University Press, 1982.

J. Katz. "Technological Change and Development in Latin America." In *Latin America and the New International Economic Order*, ed. R. Ffrench-Davis and E. Tironi. London: Macmillan, 1982.

J. Katz and B. Kosacoff. "Multinationals from Argentina." In *The New Multinationals*, edited by Lall et al.

S. Lall. *Developing Countries as Exporters of Technology: A First Look at the Indian Experience.* London: Macmillan, 1982.

——. *Developing Countries in the International Economy.* London: Macmillan, 1981.

——. "India's Technological Capacity: Effects of Trade, Industrial and Science and Technology Policies." In Fransman and King, *Technological Capability.*

——. "Multinationals and Technology Development in Host LDCs." Paper presented at the seventh world congress of the International Economics Association, Madrid, September 1983, on "Structural Change, Economic Interdependence and World Development" (transactions forthcoming).

——. "India," *World Development.* Special issue on "Exports of Technology by Newly Industrializing Countries," 12 (1984), 535-65.

S. Lall, E. Chen, J. Katz, B. Kosacoff, and A. Villela. *The New Multinationals: The Spread of Third World Enterprises.* Chichester: J. Wiley, 1983.

S. Lall and S. Mohammad. "Technological Effort and Disembodied Technology Exports: An Econometric Analysis of Inter-Industry Variations in India." *World Development* 11 (1983): 527-35.

Reserve Bank of India (Bombay). *Bulletin*, March and August 1982.

N. Rosenberg. *Perspectives on Technology.* Cambridge: Cambridge University Press, 1976.

F. Stewart. "International Technology Transfer: Issues and Policy Options." World Bank Staff Working Paper no. 344, 1979.

D. J. Teece. *The Multinational Corporation and the Resource Cost of Technology Transfer.* Cambridge, Mass.: Ballinger, 1976.

S. Teitel. "Towards an Understanding of Technical Change in Semi-Industrialized Countries." *Research Policy* 10 (1980): 127-47.

United Nations, Centre on Transnational Corporations. "Technology Transfer through Transnational Corporations in Capital Goods Manufacture in Selected Developing Countries." New York: U.N., Centre on Transnational Corporations, 1982. Mimeograph.

L. E. Westphal. "Empirical Justification for Infant Industry Protection." World Bank Staff Working Paper no. 445, 1981.

L. E. Westphal and Y. Rhee. "Korea's Revealed Comparative Advantage in Exports of Technology: An Initial Assessment." Washington, D.C.: World Bank, 1982. Mimeograph.

4

Stimulating Effective Technology Transfer: The Case of Textiles in Africa

Lynn K. Mytelka

INTRODUCTION

Textile production has historically played an important role in economic growth. It sparked the first industrial revolution, and the relatively low capital requirements, limited scale economies and simplicity of technology in this industry enhanced its role as a leading sector in other early industrialization processes as well. Even today, with the exception of man-made fibers, which are more properly classified as chemical products, the textile industry remains relatively more labor intensive than other industries, and its technology is reasonably stable, mature, and widely accessible. These characteristics, coupled with the availability of cotton, a key raw material input, and the importance of this industry in a development strategy designed to meet basic human needs, have made textiles a preeminent industry for transfer to the Third World.

The growth of a factory-based textile industry in early nineteenth century Europe and North America was largely an indigenous affair, necessitated by British refusal to undercut its market dominance in textiles by the sale of textile machinery, and aided by a flow of immigrants to the Continent and the New World.[1] Thereafter, however, the roots of modern, large-scale textile production lay not in the evolution of indigenous spinning and weaving techniques

The author wishes to thank Thomas Biersteker, Bernadette Madeuf, Peter Robson, and Larry Westphal for their helpful comments.

but in the transfer of new product, process, and organizational technologies from abroad. Consumption and production technologies, moreover, were closely linked to each other and to organizational technological forms, as Takeo Izumi observed with respect to the modern cotton spinning and weaving industries in Japan, which "did not develop from the traditional production pattern which had existed from the Tokugawa period . . . [But] were created as a result of large-scale mechanized plant transfer which in itself is a uniquely capitalistic form of production."[2]

In some of its new sites, notably Japan and, later, South Korea[3] and India,[4] large-scale, mechanized production became a dynamic element in the growth of domestic industry. Over time, textile firms in these countries became internationally competitive, and this further stimulated the growth of output and the establishment of domestic economic and technological linkages.

In other cases, as for example in much of Africa, the textile transfer process did not give rise to efficient production within a dynamic industrial sector. For unlike Japanese or South Korean firms, textile companies in Africa rarely sought to master the technology they imported and thus could not go beyond the operation of imported machinery to the modification and improvement of production processes or the design of new or allied products. This failure had remarkably little to do with Africa's material or human environment – an environment in which cotton production, textile design, and artisanal manufacture long predated that continent's colonial experience. Nor can one attribute the lack of technology mastery to an absence of sufficient trained manpower where many of Africa's textile mills are nearly a generation old. However, given the wide availability of a broad range of textile machinery on the international market,[5] the lack of a textile machinery producing sector in Africa was not a factor limiting the mastery of textile technology in the sense used above. How might we then explain the lack of technological mastery in Africa's textile firms?

In this chapter the failure to master imported textile technology is examined within the context of two of Africa's most industrialized countries, Kenya and the Ivory Coast. To widen the empirical base and thus to make generalization possible these data are then compared with case material drawn from Nigeria and Tanzania.* A

---

*In the period under consideration, the four countries chosen situated among the 43 independent countries of Africa in 1976 in the following order

total of 26 firms are thus covered in this study, of which 42 percent had been in operation for more than 15 years and a further 35 percent for at least ten years.

On the basis of these data, an attempt is made to explain the observed lack of effective technology transfer in the African textile industry and to elucidate those conditions that appear to stimulate technological assimilation within technology-importing firms. At the microeconomic level, effective technology transfer, it will be argued, is essentially a process of building indigenous technological capabilities through consciously engaging in learning-by-doing within the firm. The nature of the firm, the structure of the industry, and the policy environment, it will be suggested, are critical in stimulating this process.

EFFECTIVE TECHNOLOGY TRANSFER: DEFINITION AND MEASUREMENT

In the classical literature on technology transfer, any increment to productive capacity resulting from the import of foreign capital or machinery is regarded as a technology transfer. Yet an increment to productive capacity does not necessarily imply an increment to technological capability. Put another way, importing embodied technology (machinery and equipment) and learning to operate it does not in itself constitute a capacity to reproduce that technology or to use it for purposes other than that for which it was originally designed. For a technology transfer to be truly effective, it will be argued here, the imported technology must be assimilated, since it

with respect to levels of industrialization: measured as the proportion of gross domestic product (GDP) accounted for by manufactured value added (MVA), 19 African countries were in the less than 8 percent category, 10 fell between 8 and 12 percent, and these included Nigeria and Tanzania; a further 8 countries were in the 12-15 percent category, and 6, including the Ivory Coast (16.1 percent) and Kenya (15.5 percent), were in the over 15 percent category. It should be pointed out, however, that Nigeria's MVA accounts for 14.1 percent of total MVA of all African countries, whereas the contributions of the Ivory Coast, Kenya, and Tanzania to total African MVA each fall in the 1-4.5 percent range. In terms of the rate of growth in MVA over the period 1970-76, moreover, whereas 10 countries in the region had negative or zero rates, Tanzania, along with 14 other countries, fell in the 0-5 percent range; the Ivory Coast and Kenya (plus 9 others) in the 5-10 percent range, and Nigeria (plus 10 others) in the over 10 percent range.

is only through technological mastery that its subsequent modification, improvement, and extension becomes possible. By permitting flexibility in response to changing costs, tastes, and competitive conditions, technological mastery increases the long-term viability of the firm as well as its contribution to domestic economic growth and social welfare.

In earlier analyses of the process of technological mastery within industrial firms, the concept of learning-by-doing was frequently applied.[6] Improvements in efficiency within a plant were measured by increases in labor productivity over time, as in Lundberg's study of the Horndal steel works[7] and David's analysis of the Horndal effect in the cotton textile mills of Lowell, Massachusetts,[8] or by decreases in unit costs, as in Hollander's study of five U.S. rayon plants.[9] These increases in efficiency were then attributed, inter alia, to the automatic effects of learning in the course of production.

Recent research by Bell, Scott-Kemmis, and Satyarakwit,[10] however, revealed, in the case of a Thai galvanized steel plant that they subjected to intensive study, that there was a decline in operating efficiency over time. This and other work puts into question the extent to which, within a Third World context, changes in efficiency are an automatic consequence of participation in the production process.[11]

In opposition to earlier conceptualizations of the process of technological mastery, the approach taken here, therefore, emphasizes the dynamic and reinforcing nature of a conscious process of learning-by-doing. It does not assume that the mere "experience of production yields information and stimuli which prompt the making of improvements."[12] Rather, it is suggested that such information contributes to technological mastery only to the extent that it is related to technological knowledge about the underlying process and about the possibility that changes might be made. The connection between problem identification and problem solving thus involves a supply-side consideration, that is, it presupposes the building of indigenous technological capabilities within the firm through prior learning experiences and/or access to research and development facilities and funds outside the firm, as, for example, in state organisms, university research institutes, or supplier firms. In the former instance, trouble shooting and resolving bottlenecks in consumption, production, or organizational technologies are important means through which technological capability can cumulate within the firm.[13] In the latter, the connection between such institutes and firms is itself highly problematic.[14]

For learning-by-doing in this dynamic sense to occur, moreover, there is also a demand-side consideration that must be present. The stimuli to learning-by-doing might come from any number of sources – changes in tastes, incomes, competition, availability of inputs or foreign exchange, state policies, or others.[15] In the absence of such stimuli, however, there would be little inducement to reduce costs, improve productivity, or expand output through product, process, or organizational changes.

Taking this conceptual approach as a point of departure, it is possible to identify a series of learning experiences in the pre- and postinvestment phases of an industrial venture that contribute to the development of technological capability within the firm. The extent to which firms participate in each of these learning experiences can, in some sense, be taken as a proxy for the degree to which technological capabilities are cumulating and thus the transfer process has been effective. From this perspective, the failure of firms to undertake such steps cannot simply be ignored but must be explained.

For the technology transfer process to be effective, it is important to recognize that the process of technological mastery must begin from the moment of product conception. As Frances Stewart has pointed out, the very choice of product, given its particular characteristics, will orient choice of technique and in some instances reduce considerably the number of available technology suppliers.[16]

In order to maximize the learning potential of preinvestment activities, the recipient state or private enterprise, singly or in conjunction with industry associations or government agencies (central bank committees, royalty commissions, investment review boards) must be in a position to

1. define exactly what technology is wanted;
2. identify potential suppliers and do the research necessary to select one or a small number who can best meet these needs and with whom the best bargain can be struck;
3. prepare thoroughly for negotiations by collecting information from a variety of sources;
4. contract only for the supply of the *core* technology over which the supplier has a genuine monopoly and subcontract the *peripheral* technology to local sources or other, cheaper, foreign sources.[17]

Local firms or partners must also undertake to ensure that investments are made in the creation of central repair workshops, quality control facilities, experimental laboratories, and other loci of future innovative activity. During the investment phase, indigenous participation in design, construction, and start-up accelerates the process of mastering the use of imported production technology and permits a more rapid movement toward meeting design levels of plant efficiency and output.[18] In the case studies that follow, an attempt will be made to assess the extent to which such opportunities for learning-by-doing in the preinvestment phase have been seized.

Once the plant is built, technological changes are still possible. In fact, postinvestment modifications, improvements, and extensions to imported technology are activities that provide the kinds of learning experiences most likely to contribute to a firm's potential for flexible adjustment to internal bottlenecks or external stimuli. In undertaking such technological changes, moreover, costs may be reduced, efficiency increased, products or processes rendered more suitable to the local environment, and wider linkages established throughout the economy.

From recent studies[19] it is possible to summarize some of the many changes in consumption, production, and organizational technology that could promote technological capability within the firm and have secondary positive effects.

1. Changes in the characteristics of the product – its quality, its specifications – may open new markets and make possible new inputs.
2. Changes in the composition of inputs and their sources may reduce the import intensity of production and lower foreign exchange costs.
3. Changes in production techniques may allow cheaper material inputs to be used or may reduce the volume of material inputs required per unit of output.
4. Changes in the organization of production may increase output per man-hour or per employee.
5. Worker training may permit workers to carry out tasks more expeditiously and with greater skill and less wastage.
6. Modifications in equipment or machinery may reduce downtime, increase output per employee (capital stretching), or

improve products; it may also strengthen design capabilities and lead to machine-building capacity.

7. Changes in control systems may produce efficiency gains by generating more information about plant performance, permitting the identification of specific modifications.
8. Development of energy-saving devices or waste recycling mechanisms may reduce costs and stimulate further innovative activities within the firm or in allied industries.

The above list demonstrates the wide range of activities that could be undertaken without major new capital investments by a firm in which imported technology is progressively being assimilated, modified, and improved upon. As the technology is mastered, moreover, recourse to expatriate technical assistance will diminish. Building indigenous technological capabilities would thus serve to reduce the foreign exchange costs of technology transfer as it opened new avenues for technological innovation in the future. The extent to which learning-by-doing in this dynamic and conscious sense is being undertaken within both import-substituting and export-oriented textile firms in the Ivory Coast will be examined in the next section. Then some comparisons with import-substituting and export-oriented firms in Kenya and import-substituting firms in Nigeria and Tanzania will be offered. Finally, an attempt will be made to generalize some of these findings, to specify that set of factors that tended to reduce the incentive for learning-by-doing within these firms, thereby limiting the effectiveness of the technology transfer process, and to set forth those conditions that appear to stimulate technological mastery within the firm.

THE TRANSFER OF TEXTILE TECHNOLOGY TO THE IVORY COAST

In the first two decades following its independence, the Ivorian industrial sector grew rapidly. Industrial turnover rose from 52,407 million francs *Communauté Financière Africaine* (CFA) in 1966 to 794,768 million francs CFA in 1979-80.[20] In real terms, the rate of growth of industrial turnover averaged 11.6 percent during the decade of the 1970s.[21] As in earlier industrializers, the textile industry has played an important role in this process, and textile turnover rose from 10.3 percent of industrial turnover in 1966 to 12.5 percent in 1979-80.[22]

Three generations of textile firms are active in the Ivory Coast. Unlike those earlier industrializers in which indigenous textile entrepreneurs formed the core of a dynamic sector, in the Ivory Coast foreign capital has, since the outset, played a predominant role (Table 4.1). Of the eight firms that constituted the Ivorian textile industry in the late 1970s, Gonfreville is the oldest and most diversified. Established as a spinning mill in 1921 by a colonial official, it was taken over in the 1970s by a French commercial firm, Optorg, in partnership with the engineering subsidiary of a large French textile company, Texunion, and a set of foreign banks. This combination of foreign commercial companies, technology suppliers, and banking capital became the typical structure of ownership in the Ivorian textile industry. In each instance, moreover, the technology supplier also managed the firm.

In contrast to Gonfreville, Impressions sur Tissus de Côte d' Ivoire (ICODI), Uniwax, Société Industrielle Textile de Côte d'Ivoire (SOTEXI), and Société Ivoirienne de Textiles Artificiels et Synthétiques (SOCITAS) were set up during the 1960s primarily as import-substituting ventures by commercial firms already present in the Ivory Coast and anxious to maintain their share of the Ivorian market. It was the commercial firm that then recruited the technology supplier, which, by commercializing its technology, acquired some 10 percent of the new firm's share capital. ICODI, for example, was founded by three large commercial companies, Scoa, Cfao, and Cfci (Unilever); Schaeffer Engineering, the engineering division of a major French textile company; and a number of private banks and national aid agencies. Subsequently Riegel, the U.S. manufacturer of Wrangler blue jeans, also became a partner.

The impetus to establish final-stage manufacturing activities in the Ivory Coast came as a result of government efforts to accelerate the industrialization process. To induce foreign capital to move into industry,[23] the profitability of such ventures was ensured by a series of policy instruments implemented during the period 1958-68. These included an investment code that guaranteed industrial investments against the risk of nationalization, of nontransferability of profits or capital, and of nonconvertibility of currency. The investment code, by according a firm priority status, also provided for accelerated depreciation as well as exemptions from corporate income

TABLE 4.1

Ivorian Textile Firms: Establishment, Ownership, Size (Turnover and Employment), and Activities

	Year		*Ownership as a % of Share Capital*					
Company	*Established*	*Start-up*	*State*	*Pvt. Ivorian*	*Foreign*	*Turnover m F CFA 1977-78*	*Employment 1977-78*	*Activities*
Gonfreville	1921	n.a.	33.0	21.8	33.0	11,700	3,268	S, W, P, C
SOCITAS[a]	1966	1969	56.6	7.0	36.5	2,956	228	S, W (synthetic)
ICODI	1962	1964	31.8	–	68.2	6,100	378	P
COTIVO	1973	1976	28.6	–	71.4	4,631	1,500	S, W
SOTEXI	1966	1967	35.0	–	68.0	8,000	464	P
UTEXI	1972	1974	20.3	–	79.7	4,316	1,529	S, W
UNIWAX	1966	1970	–	15.0	85.0	7,000	666	P
SIVOITEX	1974	1975	–	7.2	92.8[b]	600	106	W, P

Key:

S = spinning
W = weaving
P = printing
C = clothing

Notes:

[a]Indented firms are affiliates of the firm preceding. Gonfreville owns 32 percent of SOCITAS, and ICODI owns 90 percent of COTIVO.

[b]Primarily locally resident Lebanese.

Source: Government and company interviews.

taxes, property taxes, and import duties on transportation equipment, raw materials, intermediates, and machinery used in the production process for up to 10 years. In addition, a *convention d'éstablissement* could be signed, thereby ensuring an enterprise of fiscal stability over a 25-year period. All eight of the textile firms have had priority status, and most are also the beneficiaries of fiscal stability agreements.

In addition to the provisions of the investment code, state lending mechanisms, such as the Banque Ivoirienne de Développement Industriel (BIDI) and the Société Nationale de Financement (SONAFI), were created to provide loans, loan guarantees and interest subsidies to industrial investors.[24] High tariffs were also maintained and indeed had been sought by foreign investors as a means to protect the domestic market. Import licensing was also practiced and in textiles virtually excluded the entry of comparable goods. Finally, a system of price fixing (*homologation des prix*) was introduced. Under it textile producers provided the state with information on their production costs, and, in concert with the state, market prices were set in function of these costs. Ivorian efforts to attract foreign investment proved so successful that by 1971, 92 percent of the capital invested in the textile industry was foreign owned.[25]

Although the first decade of independence brought about a dramatic rise in agricultural and industrial output and the Ivory Coast came to have among the highest per capita incomes in sub-Saharan Africa, by the end of the 1970s there were numerous disquieting signs that the Ivorian growth strategy was in difficulty. Incomes had increasingly become more concentrated over the decade, with the share of national income accruing to the poorest 60 percent of the population falling from 30 to 22 percent.[26] The slow growth of the domestic market that this implied threatened to undermine import-substituting activities in such consumer-oriented industries as textiles.

Intersectoral linkages established by foreign firms in the Ivory Coast were exceedingly weak, even in industries that made use of local inputs.[27] Finally, and most immediately pressing from a political perspective, was the rising rate of urban unemployment. By 1969 demonstrations of the unemployed, rent strikes, tax boycotts, and student unrest had become too prevalent to ignore,[28] and in that year President Houphouet-Boigny initiated a new economic strategy for the 1970s. It sought to encourage foreign capital in import-substituting industries to move toward export-oriented manufacturing in order to overcome the narrowness of the domestic market and the lack of backward linkages within the economy. Textiles were the keystone of this strategy. What distinguishes the third-generation textile firms Cotonniere Ivoiriene (COTIVO), Union Industrielle Textile de Côte d'Ivoire (UTEXI), and the Gonfreville *Grand Ensemble*, therefore, is that they are export oriented. COTIVO, for example, was designed to produce denim for an upstream export-oriented blue jeans manufacturing firm (Blue Bell) established by Riegel Textiles, while UTEXI and the Gonfreville *Grand Ensemble* were intended to export mass-market textiles, such as bedsheets, to Europe.

As the new economic strategy unfolded, however, it soon became apparent that new inflows of direct foreign investment (DFI) would not be sufficient to realize the Ivory Coast's industrial objectives. In 1971 and 1972 DFI amounted to only 4.4 billion francs CFA each year, whereas new investment in industry totaled 14 billion francs CFA in 1971 and 39 billion francs CFA in 1972.[29] Although DFI rose in the period after 1973, in large part this was due to a resort by foreign firms to reinvest earnings, which in the period 1973-79 amounted to over 80 percent of total foreign private investment.[30] For balance of payments reasons, moreover, the French treasury in the early 1970s restricted the outflow of capital and encouraged French firms to borrow abroad to finance their investments.

Under these conditions a more directly interventionist role for the state in the economy was thought necessary in order to sustain the growth process. In industry this was reflected in the dramatic rise in state participation in the share capital of these firms. As Table 4.1 reveals, the state now took shares in six of the eight textile firms. Yet in the context of continued reliance on foreign capital and technology suppliers, this new role for the state had little impact on the building of indigenous technological capabilities through the technology transfer process.

State intervention was primarily designed to reduce the risks to foreign investors[31] and not to influence the choice of technology or its subsequent assimilation. Indeed, technological decisions were left entirely to the technology supplier (Tables 4.2, 4.5). Their choices, however, were highly influenced by state policies. State participation in the share capital of the firm, interest subsidies and

TABLE 4.2
Nature of the Process Technology

Firm	*Source of Technology in the Development of Production Facilities*	*Choice of Technique and Modifications to the Production Process*	*Capital Intensity of Production: K/L in millions of francs CFA*[1]			
			1974	*1975*	*1976*	*1977*
Gonfreville (S, W, P, C)	Texunion – tech. supplier, partner, plant manager	*Grande Ensemble* none – open-end spinning and most modern tech. chosen Sulzer Looms	2.37	2.27	2.67	3.46[2]
SOCITAS (S, W)	Texunion – tech. supplier, partner, plant manager	None – Swiss looms, German spindles	6.23	8.05	11.48	12.33
UNIWAX (P)	Subsidiary of parent firm (Vlisco/Unilever)	None – identical to tech. supplier's core process – some greater labor intensity in peripheral tech.	–	–	–	–[3]
ICODI (P)	Schaeffer Engineering – tech. supplier, partner, plant manager	Most modern printing equipment is installed as it comes on the market – no modifications	3.75	3.69	5.07	6.09
COTIVO (S, W)	Schaeffer Engineering – tech. supplier, partner, plant manager	183 Saurer looms mischosen for denim; Ruti shuttleless looms (24); Picanol looms (432)	–	–	–	–
SOTEXI (P)	Unitika – tech. supplier, partner, plant manager	Japanese and Dutch machinery for printing, but cloth not sanforized up to export levels	3.72	3.69	3.98	4.71

UTEXI (S, W)	Unitika – tech. supplier, partner, plant manager	Conventional technology chosen: Belgian looms with shuttles and Japanese ring spinners	28.70	6.84	4.10	4.40
SIVOITEX (S, W)	Hired European textile technicians with 20 years experience to assist them	Conventional technology chosen: looms with shuttles, reconditioned spinning machinery	–	3.99	5.29	4.58

Key:

S = spinning
W = weaving
P = printing
C = clothing

Notes:

[1] K = gross cumulative capital investment; L = foremen (maitrise) + production workers (personnel qualifié 3, 4, 5)
[2] Does not include Grand Ensemble
[3] (–) data not available

Source: Government and company interviews.

loan guarantees provided by the state, and loans made available by state investment institutions all cheapened the cost of capital to foreign partners and increased the tendency for technology suppliers to choose capital intensive technology for these new plants.

Such technological choices were made not only because funds were readily available but also because the technology suppliers reaped most of their financial rewards from technology assistance payments and from commissions on the purchase of machinery and equipment that they arranged. These commissions were paid as a percentage of total investment costs (Table 4.3).

Given state policies that cheapened the cost of capital, much of this investment cost was met by borrowing. To establish ICODI, Schaeffer and its partners borrowed 49 percent of the initial investment from the BIDI and a consortium of private banks. Similarly, 71 percent of SOTEXI's investments were made with funds borrowed locally from the BIDI and a consortium of local banks, and 64 percent of the 5,600 million francs CFA invested in UTEXI were also borrowed – mostly in the form of long-term loans from the Dutch aid agency, the European Investment Bank and the BIDI.[32]

Choice of sophisticated production technology not only increased the costs of investment but also inflated the wage bill because of the need to employ larger numbers of expatriate personnel (see the Gonfreville case below). High interest charges, substantial technology payments, and the cost of expatriate labor all contributed to the relatively higher costs of production in these firms as compared with their Asian competitors. Despite this, substantial annual increases in sales were recorded, and this permitted a high rate of compensation to each of the foreign partners – the banks in interest rates, the technology supplier in technology payments, expatriate salaries, and commissions, the commercial firms in the commercialization of output within a protected domestic market in which high costs could be passed along to the consumer in higher prices – irrespective of the reported annual rates of profit (Table 4.4).

By guaranteeing the profitability of these investments through tax concessions and accelerated depreciation, which led firms to foreshorten their break-even point, through the tariff structure and through the processes of price harmonization and import licensing practices with respect to textiles, the state diminished the incentive in these firms to move toward more efficient production. Thus

TABLE 4.3
Annual Technology Payments by Ivorian Textile Firms[a]

Gonfreville	A technical assistance contract with Texunion – 25 million francs CFA per year plus 2 percent of the value of all machinery chosen by Texunion.
SOCITAS	1969-76 – a technical assistant contract with Texunion – 2.5 percent of turnover plus 5 percent of the costs of machinery, dyes, and other purchases. 1969-79 technical assistance contract with Gonfreville – 0.5 percent of turnover. 1977-present engineering contract with Texunion – "low payments"
UNIWAX	4 percent of turnover to its supplier/partner for a license to use this "secret" process, plus 1.5 percent of turnover for technical assistance and 0.5 Dutch florins per meter of cloth produced with designs from its tech. partner. In 1976-77 these technology costs amounted to 393 million francs CFA or 8.2 percent of turnover seven years after start-up. In addition it pays its parent firm 1.5 percent of turnover for administrative and fiscal services.
ICODI	A technical assistance contract with Schaeffer, its supplier/partner – 3 percent of turnover
COTIVO	5 percent of turnover
SOTEXI	5 percent of turnover until 1977, then 2 percent. Cost of Japanese personnel was 1.94 percent of turnover in 1978.
UTEXI	5 percent of turnover until 1977, then 2 percent. Cost of Japanese personnel was 4.3 percent of turnover in 1978.
SIVOITEX	5 percent of profits (not turnover) are paid to the suppliers of its spinning machinery, which was reconditioned and for which a guarantee of productivity was signed for three years.

Note:

[a]In all instances in which the partner/tech. supplier manages the plant, salaries and benefits are borne by the Ivorian firm. These additional costs are not included here.

Source: Government and company interviews.

TABLE 4.4
Ivorian Enterprise Performance Characteristics, 1973-77

Company	*Average annual change in sales (percent)*	*Average annual profitability[a] (percent)*	*Average annual exports as a percent of sales*	*Average annual change in total employment*	*Average annual percent of inputs imported[b]*
Gonfreville	25.8	6.2	39 percent to Europe and Africa	6.9	43.6
SOCITAS	38.6	11.0	5 percent to Africa only	27.0	85.8
ICODI	26.7	12.3	22 percent to Africa only	4.7	63.7
COTIVO	start-up 1976	(losses)	0 percent	start-up 1976	n.a.
SOTEXI	22.9	0.6	8.3 percent to Africa	–4.6	47.7[h]
UTEXI	24.9 (1975-77)	–6.4	40.2 percent to Europe	17.2	12.6 (1976, 1977)
UNIWAX	46.0[c]	9.0[d]	7 percent to Africa only	31.0[e]	45.7[f]
SIVOITEX	52.6 (1976-77)	–24.7[g]	7.3 percent to Africa only (1976, 1977)	–1.5	51.8 (1975-77)

Notes:

[a]Profitability is measured as the average "net result" as a percentage of turnover in 1973-77.

[b]Imported inputs as a percentage of total purchases (average annual for the period 1973-77).

[c]But quantity increased by only 16 percent. The increase in sales reflects price increases in cloth.

[d]Declared after tax profits as a percentage of turnover.

[e]The figure represents 1977/1972 data.

[f]Imports as a percentage of the value of the final product.

[g]Few of the firms in this survey were subject to corporate profits taxes. SIVOITEX, however, was and this might have led to a deflation of "gross results" in order to reduce taxable income.

[h]Before Utexi started-up 80 percent of total purchases were imported. This fell to 31 percent in 1976 and 1977.

Source: Government and company interviews.

changes in the productivity of labor and rates of capacity utilization, where data were available, were mostly low (Table 4.5).

In addition to the absence of process modifications aimed at increasing productivity, Table 4.4, column 5 illustrates the lack of change in the source or amount of imported inputs used by these firms. Product changes undertaken in this period, moreover, were largely confined to the introduction of polyester-cotton blends in imitation of parent company/country taste shifts. The decision to move in this direction, however, increased the import intensity of production, engendering both a need for new machinery and a continuous stream of imported intermediate inputs thereafter.

Finally, as Table 4.5 reveals, all product and process modifications were conceived and implemented by the technology supplier/partner. Only a limited Ivorianization of managers and technicians has occurred (Table 4.6). The little learning-by-doing that did occur in the course of production thus cumulated in expatriates and not Ivorians. There are no cases in the Ivory Coast of technical or managerial personnel trained in these firms transferring this knowledge to independent enterprises, nor are there examples of Ivorian-managed textile firms from among the foreign affiliated companies.

Only Société Ivoirienne de Textiles (SIVOITEX), owned by resident and foreign Lebanese individuals and banks, did not rely exclusively on a foreign multinational partner for its technology and its management. Both product and process choices within SIVOITEX reflected the more domestic-market orientation of this ownership structure. Initially established to meet a demand for towels in the local market, the company chose to use "waste" or "regenerated" yarn rather than new material in its product. As the market developed, to reduce the cost of imported yarn, it integrated backward and chose reconditioned spinning machinery in order to do so. These choices were made by the local management in consultation with European technicians and machinery suppliers. Product modification to improve quality, research into designs and colors, and efforts to reduce costs have characterized the operation of this firm in its first four years of production.

As relatively inefficient producers, the Ivorian foreign textile affiliates needed the margin of protection provided by the Lomé Convention or the General Systems of Preferences if they were to fulfill their export mandate. These firms thus became increasingly vulnerable to even small shifts in demand or in the level of

TABLE 4.5

Technological Capability and Technological Change in Ivorian Textile Firms

Company	*Present tech. links abroad*	*Recent or planned new products*	*Source of tech. for new products/ or processes*	*In-house research and development activities*	*In-house repairs*	*Changes in productivity (O/L)*
Gonfreville	Tech. Asst. contract/ Texunion	yes – cotton & polyester mixtures & printing	selected by Texunion – all machinery imported	Quality control lab & 4 part-time designers for cloth printing	yes	n.a.
UNIWAX	License from Parent affiliate (Unilever-Vlisco)	none – though designs are more sophisticated	Licensee – all machinery is imported	none – some design capability	some	a total of +3% over the period 1972-77
SOCITAS	Tech. Asst. contract replaced by engineering contract with Texunion in 1976.	cloth for woman's wear; upholstery & velours	Texunion employees at SOCITAS plus info. from machinery suppliers. Machinery imported	some R & D on dyeing with help from dye companies	none	n.a. capacity utilization 1977-78 88.9%
ICODI	Schaeffer engineering	polyester & cotton begun in 1975-76	Schaeffer – all machinery is imported	no R & D though Production Mgr. and 4 technicians adjust process to new designs	yes	+79% 1972/1976 –97% 1977/1979

COTIVO	Tech. Asst. Schaeffer	no	n.a.	They spent 1% of turnover improving denim quality – with 3 part-time professionals	yes	negative change as Saurer looms are converted to other uses
SOTEXI	Unitika Tech. Asst.	originally 100% cotton "pagnes" now 20% dress, & shirt fabric; moving to 15% polyester	Unitika-employees	Does R & D on colors & chemicals in their labs Do own quality control	yes except electric	capacity utilization 1977-78 60%
UTEXI	Unitika	at first 100% cotton, in 1978 began polyester	Unitika-employees (polyester imported from Unitika)	no R & D; do quality control	some	capacity utilization 1980 = 48.3%
SIVOITEX	European technicians	have diversified from towels to covers & are testing upholstery fabrics	Imported machinery	No specific R & D costs but 1 person half-time from Beaux Arts de Paris is doing design & color research. Do small modifications to machinery.	yes	n.a.

Note: Output in meters or tons/production workers.

Source: Government and company interviews for all but capacity utilization figures which were calculated from Ediafric (1982), p. 40.

TABLE 4.6
Ivorianization of Cadres[a]

	Ivorian cadres as a proportion of total				
	1974	1975	1976	1977	Training Programs
Gonfreville	4/28	5/24	12/41	15/49	No explicit programs
SOCITAS	0/3	1/4	1/6	1/6	A small number of foremen have done short-term (two-week) training programs in unaffiliated European firms.
UNIWAX	—[b]	—	—	7/14	All workers were trained on the job.
ICODI	4/12	5/12	7/14	7/15	Some Ivorian technicians were sent to Europe for training.
COTIVO	—[c]	—[c]	—[c]	—[c]	Some in-house training. Most of the workforce has no factory experience. A training center with two Picanol and two Saurer looms was established.
SOTEXI	1/27	1/24	2/27	3/26	On-the-job training provided.
UTEXI	3/29	3/31	3/29	4/33	On-the-job training provided.
SIVOITEX	—	0/3	0/2	0/5	None

Notes:

[a]Includes technical and administrative managers.

[b]In 1970 there were 24 expatriate cadres and foremen and no Ivorians. By 1977-78 there were 7 expatriate cadres and 9 expatriate foremen and 7 Ivorian cadres and 10 Ivorian foremen and office staff. Three of the 4 textile engineers are Ivorian.

[c]COTIVO started up in 1976 with 27 expatriate cadres and foremen. There were still no Ivorian cadres or foremen in 1979, although 7 are being trained.

Source: Government and company interviews.

competition within the Western European market. When neoprotectionist pressures led to revisions of the Lomé Convention that eliminated their advantage,[33] the lack of Ivorian technological mastery made adjustment to these changes costly and difficult, as the case studies below illustrate.

COTIVO

In 1970, Mohamed Diawara, then minister of plan, approached ICODI with a view to persuading its owners to integrate backward toward spinning and weaving in keeping with the newly adopted export-oriented growth strategy. As ICODI's priority status was soon to expire and the firm would no longer be permitted to import free of customs duties the greycloth it printed, backward integration was now in its interest. The state agreed to grant the new firm, COTIVO, priority status and took an equity share of 28.6 percent. Schaeffer, which managed ICODI, would manage COTIVO and assume all responsibility for the choice of technology. The state did not participate in the choice of technique, the identification of potential suppliers, or negotiations over the terms of technology transfer. There was no attempt to depackage the technology, and no Ivorians were associated with Schaeffer engineering in the design or production engineering phases.

COTIVO, established in 1973, was intended to produce greycloth for both Uniwax (owned 85 percent by Unilever, which is also a shareholder in ICODI and COTIVO) and ICODI's printing and finishing plant, thus ensuring that the spinning and weaving operations would be able to take advantage of economies of scale. It was also designed to produce denim for Riegel's new blue jeans subsidiary in the Ivory Coast.

COTIVO received a 1,500 million francs CFA loan from the European Investment Bank and additional long-term loans amounting to some 1,700 million francs CFA from the BIDI, the West German aid agency, and other national and international public lending agencies. In addition, COTIVO had a medium-term loan amounting to 3,200 million francs CFA from a banking consortium led by the Ivorian affiliate of the Banque Internationale pour le Commerce et l'Industrie de la Côte D'Ivoire (BICICI). Loans constituted 67.2 percent of the capital invested, and financial charges ran at 15 percent of turnover in 1979. In 1980-81 the BIDI was obliged to open a further line of short-term credit amounting to 135 million francs CFA for COTIVO.[34]

Not only is COTIVO heavily in debt and thus vulnerable to shifts in demand for blue jeans and the imposition of voluntary export restraints by importing countries, but also the choice of both product and technology can be questioned. Denim is a product with high technical specifications. It permits only very limited tolerances in dyeing, streaking, and twisting. Although COTIVO's quality has improved, most denim firms operate with a 2-3 percent "second choice" whereas COTIVO's output, several years after start-up, still included 13 percent or more "second choice." In addition to high quality standards, denim production requires high volume and low cost to compete efficiently with Asian producers, who currently set world prices. COTIVO cannot meet these conditions.

COTIVO is also plagued by a number of inappropriate technological choices. The Saurer looms initially purchased by Schaeffer are not appropriate for heavy denim cloth, and they have now been converted to the production of cotton greycloth for sale to ICODI and UNIWAX. But the use of expensive looms for such a simple product pushes the cost of greycloth well above that of imports and increases the price of printed cloth produced by these two firms. UNIWAX, which has traditionally exported its "Indonesian 'Dutch'" style wax prints to neighboring West African countries, has been particularly affected, and initiated talks with the government about importing cheaper Asian greycloth rather than fulfilling its commitment to use COTIVO's output.

COTIVO's denim spinners and other related equipment, moreover, are designed to produce enough input to weave 9 million yards of cloth, but COTIVO is producing only 2.3 million yards of cloth. Hence all spinners operated only two days per week and indigo baths similarly functioned at less than one-third capacity. This further increased the cost of producing denim. As Blue Bell encountered difficulties in selling its output in Europe, it cut back production, creating further difficulties at COTIVO in the early 1980s.

Gonfreville

The combination of a limited domestic market, given the nature of product choices made in the 1970s, high indebtedness related to the type of process choices, and a lack of technological mastery by Ivorians, helps explain Gonfreville's high costs of production and the difficulties it encountered in reorienting production toward sales in Europe. At independence, Gonfreville was still quite small, with 11,000 spindles, 32 looms, and a small dyeing works. By 1969, under the impetus of Optorg and Texunion, a new printing plant was built, the number of spindles installed rose to 16,700, over 400 looms were in operation, and the enterprise employed 2,500 persons.

Some 40 percent of Gonfreville's output during the early 1960s was exported to francophone Africa. As other African countries developed domestic cotton textile industries, these markets began to close to Gonfreville's exports. The decisions to move toward the production of polyester-cotton blends and to invest in neighboring countries were taken in this context. By 1969 Gonfreville's synthetic textile spinning subsidiary, SOCITAS, and its joint venture with the government of Upper Volta, the Société Voltaïque de Textile (VOLTEX), were in production. Gonfreville was not only the first wholly integrated spinning, weaving, and finishing company in the Ivory Coast, but it was one of the largest manufacturing firms in Africa.

The emergence of competition within the Ivory Coast, however, soon slowed the growth of Gonfreville. In response a decision was taken to build the *Grand Ensemble*. Making use of the most modern open-end spinning process and costly Sulzer shuttleless looms, the *Grand Ensemble* was designed to produce for the European market. In 1971 the Ivorian state granted Gonfreville priority status in order that the tax and tariff payments saved might be accumulated by the firm and reinvested in the planned export-oriented expansion. In addition, the state purchased a 33 percent share in Gonfreville, and, like UTEXI and COTIVO, Gonfreville borrowed heavily to finance its investment program. Of the 4,500 million francs CFA invested in the *Grand Ensemble*, 1,300 million francs CFA came from the funds accumulated through the tax and tariff benefits granted by the state, and a further 2,400 million francs CFA came from a medium-term loan guaranteed by the state and provided by a consortium of foreign banks located in the Ivory Coast.

Given its high gearing ratio, financial charges for Gonfreville were quite heavy, amounting to 8.3 percent of turnover in 1977-78. High salaries and benefits were another factor in its high costs of production. In 1977, salaries and benefits amounted to 24.4 percent of production costs. Because so little effort had been expended to train Ivorians for managerial and technical positions, in 1977 only 9 of the 24 administrative and commercial cadres and 6 of the 25 technicians and technical cadres (engineers) were Ivorians. Non-Africans in these job categories earned 221 percent more per year than their Ivorian counterparts. As non-African salaries and benefits constituted 19.8 percent of the total wage bill, simply replacing Europeans with Ivorians could have reduced production costs by 6 percent.

The high import content of production was still another negative consequence of Gonfreville's technological choices in the 1970s. As the product mix shifted toward imported polyester and cotton blends and as imports of heavy cotton cloth (cretonne) increased for its finishing operations, net value added fell from 50 percent of the value of total output in 1973 to 43.2 percent in 1975 and 39.2 percent in 1977.

By 1980 a new technical imbalance had emerged between the original Gonfreville production units and the *Grand Ensemble.* In the former complex, using conventional technology, over 2,000 persons were employed in spinning and weaving operations alone. These 39,776 spindles and 593 looms had the capacity to produce 4,500 tons of cotton yarn, 300 tons of polyester yarn and 15,600 meters of cloth per year. In the *Grand Ensemble*, with the most modern technology and only 450 persons employed, 11,088 spindles, 35 shuttleless looms, and 120 conventional looms had a production of 2,800 tons of cotton yarn and 5,000 meters of cloth.

The high cost of production and the protectionist tendencies of the European Economic Community (EEC) countries during the late 1970s made it difficult for Gonfreville to market its output abroad. In 1979 many of the Sulzer looms were converted to the production of polyester shirt fabric for the domestic market, where tastes and incomes do not require production to such high technical standards. As in COTIVO, the high cost of production was covered by the process of price harmonization, which permits these high costs to be passed along to the Ivorian consumer. In 1979 and again in 1980, a marked worsening in the Ivorian balance of payments occurred. The payments crisis led to the imposition of austerity measures recommended by the International Monetary Fund (IMF). These tended to accentuate the already declining rate of final domestic demand. Real consumption of households, which had increased by 8.5 percent in 1979 over the previous year, rose only 4 percent in 1980 and fell in 1981.[35] As the government itself noted, this contraction in domestic demand further exacerbated the difficulties faced by the Ivory Coast's import-substituting industries. Within Gonfreville this has led to the phasing out[36] of operations in its original production units, which, by comparison with the *Grand Ensemble*, whose investment costs must be met by high-capacity utilization, appear inefficient.[37]

The Ivorian situation is not unique in Africa, and the combination of factors that brought it about is replicated throughout the continent. Case material from Kenya, Nigeria, and Tanzania will illustrate this point.

SOME COMPARATIVE CASES IN THE TRANSFER OF TEXTILE TECHNOLOGY

Unlike that of the Ivory Coast, the Kenyan textile industry contains both foreign and domestically owned private textile firms. In the latter case, however, indigenous owners are drawn from the resident Asian and not the African community. Large-scale African capital, rather than striking off on its own, has remained "firmly oriented to lucrative joint ventures in import reproduction" such as the Chui Soap Factory, manufacturers of high-quality toilet soap and detergents, or Tiger Shoes.[38] In contrast to some of the local Asian capitalists, African capitalists have tended to move their firms "very much in the direction of the employment-minimizing, linkage-limiting style of the transnational subsidiaries in Kenya."[39]

In addition to participation by locally resident Asians, the Kenyan textile industry is distinguished by the fact that during the 1960s and 1970s, as textile production expanded, the enterprises leading this expansion were Japanese and Indian multinational corporations (MNCs) as opposed to Western European firms. Not until the mid-1970s did a number of Western European MNCs enter the Kenyan market with a view to establishing export-oriented textile production. In several of these cases, the firms, nearing collapse, were taken over by the state. Government intervention in the textile industry consists solely, in fact, of takeovers of foreign subsidiaries in danger of bankruptcy.

With a questionnaire identical to that employed in the Ivorian study, the data presented below were collected by Steven Langdon during several research trips to Kenya in the late 1970s. They cover ten textile manufacturers – all of the larger weavers and spinners in the country. These ten firms employ some 10,373 persons, slightly more than the 8,139 employed by the eight firms that comprise the Ivorian textile industry. Installed capacity, however, is somewhat smaller in the Kenyan case. In 1980 there were a total of 85,400 spindles and 1,981 looms in Kenya, whereas three years earlier 101,076 spindles and 2,006 looms had already been installed in the Ivory Coast. Total sales turnover in the Kenya industry amounted to 839 KShs in 1980 (see Table 4.7).

In the Ivory Coast, as we saw in the previous section, there were no indigenous privately owned and managed textile firms. In Kenya, where there are four such firms, it is thus possible for the first time to assess differences in the extent to which the building of indigenous technological capabilities is affected by the ownership or management structure of the firm.

Table 4.8 provides data on the nature of technological capabilities and technological change in Kenyan textile firms. These data suggest the emergence of technological capabilities in locally owned firms. In these firms the main source of technical know-how was the knowledge and experience of their own personnel, and no institutional links were maintained with foreign technology suppliers. In all four cases new products had recently been introduced, or there were plans to do so. While the level of indigenous technological capability in state firms was not as high, it was clearly changing. S2 planned to phase out a technology contract it had retained with its former parent firm, and three of the four state-owned firms indicated that they could initiate new products on their own. In both foreign-owned firms, on the other hand, the main source of technology was the parent firm, and there was no stated intention to introduce new products designed expressly for the Kenyan market.

In earlier work on licensing and technological dependence in the Andean metalworking and chemical industries, I have argued that prior reliance on licensing as a vehicle for obtaining technology, when accompanied by an absence of efforts to master the imported technology through a conscious process of learning-by-doing, led to continued reliance on imported technology in the introduction of new or allied products later.[40] Data from the Ivorian and Kenyan

TABLE 4.7
Kenyan Textile Spinners and Weavers, 1980

	Established as Foreign Subsidiaries in the 1960s	*Private Domestic Firms*	*Newly Established European Subsidiaries in the 1970s*
Number of firms and locations	3 (2 in Thika, 1 in Kisumu)	4 (Nairobi, Thika, Hombasa, and Kiambaa)	3 (Nairobi, Nanyuki and Eldoret)
Total sales turnover (KShs.)	251 million	380 million	208 million
Employment	3,730	4,300	2,343
Looms in use	880	508	593
Spindles in use	24,000	40,200	21,168*
Products manufactured and sold	25-26 million meters of woven fabric (12 million cotton; 7 million blends; 6-7 million synthetics and nylon); some cotton yarn	19-21 million meters of woven fabric (11-12 million cotton; 7-9 million blends and synthetics); 1 million kg. of yarn; 3-4 million meters of knitted fabrics; some garments	18-19 million meters of woven fabric (13-14 million cotton) 5 million synthetic thread and yarn

*Covers only 2 of the 3 firms.

Source: Company interviews, Kenya, 1980; Interview with Industrial Adviser, Ministry of Industry, Nairobi, 1980. As reported in Steven Langdon, "Indigenous Technological Capacity in Africa: The Case of Textiles and Wood Products in Kenya." Prepared for the International Workshop on Facilitating Indigenous Technological Capacity, Centre of African Studies, Edinburgh University, U.K., May 1982, Table 1.

TABLE 4.8
Technological Capability and Technological Change in Kenyan Textile Firms

Enterprise (location, when first producing)	*Main source of Technology in Development of Production Facilities*	*Present Institutional Technology Links Abroad*	*Recent or Planned New Products*	*Main Source of Technology for any New Products/Processes*	*Present R & D Efforts Inside Firm*
Foreign Subsidiaries					
FT1 (Thika, 1963)	Japanese Parent	Technicians from Parent	No	Japanese Parent	None
FT2 (Kisumu, 1965)	Indian Parent	Technology contract with parent; license with input supplier; consultant agreement with British mnc	No	Indian Parent	None
State Firms – All Previous Subsidiaries					
S1 (Nairobi, 1973)	Former German Parent	Technology/Management contract with British mnc; trademark agreement with same firm	Yes (continuous filament synthetics)	British Management/ Technical Contract mnc	None
S2 (Thika, 1965)	Former Japanese Parent	Technology contract with former parent, being phased out	Yes (new widths and fabrics)	Own R & D lab	4 chemists in own lab; 8% of turnover
S3 (Nanyuki, 1976)	Former British Parent	None; expatriate management	Yes (garments)	Own enterprise	Beginning stress on this, via Indian processing manager

S4 (Eldoret, 1977)	Former German Parent	Licensing agreement for Sanforizing; license with Dutch mnc re designs	No	Own enterprise	None
Local Enterprises					
L1 (Thika, 1958)	Own enterprise	None	Yes (Polyester cloth)	Own enterprise	None
L2 (Nairobi, 1963)	Own enterprise	None	Yes (Filament yarns)	Own enterprise	No formal unit; only designers
L3 (Mombasa, 1964)	Own enterprise	None	Yes (Polyester cloth)	Own enterprise	None
L4 (Kiambaa, 1964)	Own enterprise	None	Yes (Printed cloth)	Own enterprise	None

Source: Langdon (1983), Table 3.

textile industries lends additional support to this hypothesis and points to the pivotal importance of ownership/management structure in this process – not because a local person is somehow more adept, but because local owners/managers respond differently to domestic environmental incentives and constraints than do multinational owners/managers.

Thus in the Ivorian case, with the exception of SIVOITEX, which had introduced a variety of new products on its own – some of which, as in the case of yarn, had required new machinery imports – the decision in textile firms to diversify product lines was made by expatriate technology suppliers/managers much as the initial conceptualization and execution had been. Similarly in the Kenyan case, firms FT1, FT2, S1, and S4, which had grown on the basis of foreign technology inputs and continued to have institutional links with foreign technology suppliers, felt capable of initiating new processes or products only on the basis of further foreign technology inputs or, as in the case of S4, "planned no such initiatives in any event" (because of its desperate financial situation).[41] In contrast, three of the four firms that had developed on the basis of their own technology planned to introduce or had introduced new products and had done so through their own technological efforts.

Table 4.9 relates the development of indigenous technological capability (ITC) to performance characteristics of these firms. Average annual increases in sales and employment were larger in textile firms that had developed some ITC, indicating that they were growing somewhat faster in the Kenyan market. Such firms also made far more forward and backward linkage investments within the economy, though, surprisingly, the percentage of imported inputs was roughly the same as that of the foreign-owned, non-ITC firms. Nonetheless, these figures do suggest a positive impact of ITC on enterprise performance.

> The . . . difference is even more dramatic when the performance of textile firms S2, S3 and S4 is considered – all of them dependent on foreign technology in their establishment and development. S3 and S4, in particular, were colossal disasters. S3 was based on machinery sold by one of its parents from an operation in Germany that the parent was shutting down – and never achieved operational viability; the enterprise was bankrupt less than two years after start up and had to be rescued by the state at a cost of some 70 million shillings (US $7 million). S4

TABLE 4.9
Textile Enterprise Performance in Kenya

	ITC Firms			
	L1	*L2*	*L3*	*Average*
Average annual sales increase	25% (1975-80)	42% (1976-80)	40% (1979-80)	36%
Average annual after tax profitability (as percent of turnover)	7.25% (1975-80)	"good"	9% (1975, 1980)	8%
1980 exports as percent of sales	0	11%	0	4%
Average annual change in employment	13% (1975-80)	15% (1976-80)	9% (1976-80)	12%
1980 percent of inputs imported	30%	60%	40-50%	45%
Linkage investments made	No	Garments, polyester filament, yarn, towels, carpets		

	Non-ITC Firms			
	FT1	*FT2*	*S1*	*Average*
Average annual sales increase	10% (1975-80)	22% (1975-80)	26% (1975-80)	19%
Average annual after tax profitability (as percent of turnover)	3-4% (1975-80)	1.25% (1975-80)	−3.4% (1977-80)	1%
1980 exports as percent of sales	0	0.1%	5.9%	2%
Average annual change in employment	4% (1972-80)	12% (1975-80)	8% (1977-80)	8%
1980 percent of inputs imported	25-30%	63%	42%	44%
Linkage investments made	No	No	No	

Source: Langdon (1982), Table 4.

> was sold very expensive equipment, and provided a high return to its parent for inputs of technological knowledge, but also lost so much money (100 million shillings) that the state had to take it over in 1980. S2, meanwhile, had provided profits to its parents on input and machinery sales, but made minimal declared profits until it too was rescued by the state in late 1978. Technological dependence on foreign enterprise, then, had been extremely costly in the textile industry.[42]

How had L1, L2, and L3 acquired the technological capability that appears to have had such a positive impact on enterprise performance? Langdon stresses two factors, in particular, in answering this question: first, that these were family-owned businesses that began small and expanded as knowledge and experience grew; second, that two had begun as small-scale garment manufacturers, later adding second-hand spinning machinery in one case and second-hand knitting machinery in the other as their respective first expansions. "These seemed clear-cut cases of learning-by-doing, of the gradual accumulation of technological knowledge as enterprises expanded and entrepreneurial experience deepened."[43] Although these local firms do make use of expatriate expertise in various functions within the firm, what distinguishes them from the foreign-owned firms is how they use it and where the locus of technological and economic decision making is situated. In the Kenyan case, Langdon's work gives us the impression that the expatriates are supplements or complements to but not replacements for indigenous technical personnel and decision-making capabilities.

Langdon's analysis of the growth of indigenous technological capabilities in locally owned enterprises through the transfer of technology suggests the existence of a number of conditions that are absent in the Ivorian case but present here and more conducive to effective technology transfer. Let us look at these locally owned enterprises a bit more closely. As family-owned businesses, these firms tend to be attuned more to local market conditions than to world market product specifications. Their initial product choices are thus different from those of a local entrepreneur who develops production through a joint venture based on previously imported products or of a foreign-owned firm seeking to reproduce its product line in the host country.[44] As family-owned businesses, moreover, such firms would probably have accepted lower price-cost differences – a factor of some importance, for example, in Argentina's family-owned machine tool companies that permitted them to survive the credit squeeze of the 1970s.[45] As members of the Asian minority in Kenya, the owners of these firms did not have the political clout needed to force an alliance with the state and thereby secure protection of the domestic market for their product or state subsidies and credits for the initial stage of production, as the MNCs in Kenya had been able to do. Such firms, therefore, were obliged, from the outset, to be cost-competitive. They could not compete on the basis of product quality or frequent differentiations in product characteristics. Both product choice and the nature of the competitive situation in which they found themselves were thus conducive to process choices that involved less capital intensity and technological sophistication than oligopolistic market competition and greater access to the state and its resources might have permitted. In turn, this rendered the imported technology more accessible to local owners and more easily mastered. Over time such firms could grow and deepen their technological know-how, and, indeed, the emphasis on cost-competitiveness implied that such a process of technological apprenticeship would involve the kinds of innovations in organizational, product, and process technologies that would permit these firms to compete successfully against multinational firms within the same industry.

Few of these conditions were met by the Nigerian textile industry as it developed during the 1960s. Established prior to independence, Nigeria's first two cotton textile mills (L-1 and S-1 in Table 4.10) were both locally owned but foreign managed. Although L-1 disappeared in the late 1970s, S-1 has since become Nigeria's largest state-owned textile enterprise.

Not until 1962 did the first multinational firm (F-3) begin textile production in Nigeria. By 1975, however, foreign-owned firms accounted for all but one of the seven largest textile companies, that is, companies with over 2,000 employees.[46] The foreign-owned sector includes, moreover, Nigeria's largest textile mill (F-4) as well as the main integrated textile companies (F-2 and F-3).[47]

Despite this inauspicious start, as reflected in the failure to adapt imported technology to local conditions, the protected nature of the Nigerian market, and the growing dominance of the textile industry by the more capital-intensive, foreign-owned firms, competitive conditions within the Nigerian market might conceivably have stimulated a process of technological change within these firms.

Little work has been done, however, on this, and the data are thus scanty at best. Nonetheless, data collected from a small sample of the largest foreign-, state-, and locally-owned textile firms (Table 4.10) suggest that this has not been the case. Table 4.10 provides data for one measure of the extent to which a process of technological apprenticeship has begun – the potential that minor process or organizational changes have to increase the productivity of labor without incurring additional capital expenditures. Capital stretching of this sort was, for example, a key element in the assimilation of imported technology by the Brazilian steel firm, USIMINAS.[48] Whereas it might be expected that a firm would move toward greater labor productivity in the post-start-up period, Nigerian firms, with rare exception, exhibit a deterioration in the volume of textile output per worker (Table 4.10). Indeed, most of the very few increases in labor productivity recorded for these seven firms over this eight-year period are associated with rising levels of capital investment in those years.* During the period 1965-70, the index of consumer prices (1975 = 100) rose from 40.1 to 52.8, but the value of textile output by these firms rose far more rapidly and, in fact, outstripped the rate of increase in the volume of textile output as well. In the firm F-4, for example, the volume of output rose by 35 percent in 1969 over 1968, but the value of goods produced increased by 54 percent. Labor productivity, on the other hand, fell by 36.2 percent. Rising textile prices in the 1960s thus permitted profits to be generated in an expanding market whose demand was satisfied by the establishment of new firms and the enlargement of existing capacity. Under these conditions, however, there was no penalty for a fall in productivity and hence little incentive to innovate.

The effect of this failure to initiate a process of technological apprenticeship was immediately apparent in the wake of the Nigerian government's decision to liberalize tariffs on textile imports as an anti-inflation measure. Output in these textile firms fell dramatically in 1972 as cheaper imports flooded the market. The response to this crisis was twofold. On the one hand, the industry pressured the government for assistance, and some measures to aid the ailing industry were incorporated in the 1973 budget. On the other hand,

*Another possibility, of course, is that changes in productivity may be tracking unobserved changes in demand for textile products.

TABLE 4.10
Nigerian Textile Firms: Changes in Labor Productivity[a]

Firm	Year of Start-up	1970[b] Activities	1970[b] Output '000 yds.	Changes in Labor Productivity 1965/1964	1966/1965	1967/1966	1968/1967	1969/1968	1970/1969	1971/1970	1972/1971
F-1[c]	1967	P	34,454	–	–	–	37.3	–14.6	–17.4	–24.7	–51.9
F-2	1965	S, W, P, K	59,525	–	–26.9	–47.9	152.0	–9.0	8.5	–17.1	–9.3
F-3	1962	S, W, P, K	45,288	–9.8	19.0	–8.3	20.3	3.5	0.0	–14.0	–42.1
F-4	1965	S, W, P	71,144	–	28.0	–5.5	–32.6	–36.2	–9.7	–31.3	–19.6
S-1[d]	1957	S, W	52,177	37.7	3.7	11.7	1.5	–2.4	–10.8	51.4	–20.8
S-2	1963	S, W	8,622	n.a.	n.a.[e]	–23.5	2.7	–8.5	–7.8	–15.8	n.a.
L-1[f]	1951[g]	W	1,306[h]	–22.5	–16.0	n.a.	n.a.	n.a.	n.a.	n.a.	n.a.

Key:

S = spinning
W = weaving
P = printing & finishing
K = knitting

Notes:

[a]Labor productivity is measured as output in yards per operative.
[b]1970 was chosen rather than 1972, as the latter was a crisis year in which output in most factories dropped significantly.

Notes, continued:

[c]F-1, F-2, F-3 and F-4 are majority foreign-owned firms in which the Nigerian federal or state governments and/or Nigerian individuals own a minority of the share capital, generally under 25 percent.

[d]S-1 and S-2 are state-owned firms. In the case of S-2 the foreign owners sold out to the state in 1965 following a major fire.

[e]Fire closed the mill for five months in 1965.

[f]A locally owned private firm, which began as a trading company and went out of business in the late 1970s.

[g]Year of establishment as opposed to start-up.

[h]1966 data.

Source: Federal Republic of Nigeria, Federal Office of Statistics, Industrial Survey Data.

these firms now adopted a new growth strategy – one based not on technical improvements but on changes in product mix. A marked shift occurred in this period toward the production of higher-priced polyester fabrics (F-2 and F-4) and knitted goods (F-2 and F-3). In the case of the firm F-2, for example, whereas the volume of output rose from 42 million yards of cloth in 1968 to 60 million in 1970, it dropped to 46 million in 1972, to 44 million in 1974 and to 38 million in 1980, yet the value of goods produced per yard, which had risen from N£ 0.13 in 1968 to N£ 0.17 in 1970 (Nigerian pounds), remained at that level through 1974, rising dramatically to N£ 0.53 in 1980 with the introduction of new synthetics and knitted goods. Over this period, moreover, the output per worker in F-2 fell steadily from 19,000 yards in 1968 to a low of 11,220 in 1980. Although 1972 was a crisis year, this firm recorded after-tax profits in both 1972 and 1973. Foreign-owned firms thus have little incentive to embark upon a process of technological apprenticeship where profitability can be sustained by competition based on product differentiation and a shift toward higher-priced goods. Under these conditions and despite the increased Nigerianization of supervisory and managerial staff (Table 4.11) and increased indigenization of ownership consequent upon the adoption of the Nigerian Enterprises Promotion Decrees of 1972 and 1977, little change in technological

TABLE 4.11
Nigerianization of Supervisory and Managerial Staff, 1967-80

	Percentage of total supervisory and managerial staff			
Firm	*1967-69 Avg.*	*1970-72 Avg.*	*1974*	*1980*
F-1	10.3	45.1	63.6	73.8
F-2	3.3	23.7	73.6	82.8
F-3	33.7	74.7	n.a.	n.a.
F-4	7.1	37.2	85.1	76.7
S-1	7.8	26.3	n.a.	n.a.
S-2	37.6	57.0	n.a.	n.a.

Source: Federal Republic of Nigeria, Federal Office of Statistics, Industrial Survey Data.

behavior would be anticipated.* Unfortunately, a lack of data on state and local privately owned firms precludes any comparison with the Kenyan material presented earlier. The Nigerian case thus warrants considerably more investigation before any firm conclusions can be drawn.

To this point it has been argued that a process of technological apprenticeship appears to be triggered when persons in key decision-making positions within an enterprise, responsive to their domestic environment and sufficiently aware of technical possibilities, are confronted with pressures to reduce production costs. Such pressures may emanate from competitive conditions within the primary market of sale, as in the case of Kenya's Asian textile manufacturers, or, as Nathan Rosenberg[49] has suggested, may take the form of either bottlenecks in the production process resulting from internal imbalances there or changes in external conditions. The case of Kilimanjaro Textiles (KILTEX) lends additional support to this hypothesis and permits a further elaboration of these external conditions.

KILTEX, located in Arusha, Tanzania, was established in 1967 and placed under TEXUNION (France) management. It was initially foreign owned, but a majority of its shares were purchased by the Tanzanian government in 1972.

The choice of technique, in this instance the decision to purchase second-hand textile machinery, was made by TEXUNION. But the choice was an unfortunate one as frequent breakdowns resulted from a failure to ensure proper maintenance – a failure attributed to a lack of experience with this type of technology within KILTEX.[50] Although TEXUNION initiated some on-the-job training programs for mechanics, classes were affected by the continual rotation of expatriate staff. Senior positions, moreover, remained entirely the preserve of TEXUNION personnel, and this included both the design department, responsible for the design and modification of products, and the engineering department, which serviced and maintained

---

*Recent studies of firms that have become major nationally owned show no change at all in technological behavior and remarkably little participation by new national partners in the technology or investment decision-making process. Thomas J. Biersteker, *Distortion of Development? Contending Perspectives on the Multinational Corporation* (Cambridge, Mass.: MIT Press, 1978); Ankie Hoogvelt, "Indigenization and Foreign Capital: Industrialization in Nigeria," *Review of African Political Economy* 14 (January-April, 1979): 56-68.

the machinery, ensured quality control, and was responsible for modifications to machinery and processes required in the course of production.

In 1974, disappointed with the pace of Tanzanianization within KILTEX, the government, at the expiration of TEXUNION's contract, chose the Pakistani consulting firm of Seigo Brothers as the new managing agents and assigned them the specific task of training Tanzanian replacements.[51] Within a few years Tanzanian engineers and technicians had taken charge of the engineering department. As the Tanzanian balance of payments position deteriorated in the 1970s, following two dramatic oil price increases, and foreign exchange became scarce, the engineering department began to make rapid progress in the manufacture of spare parts.

> All told, 75-85 percent of the spares are manufactured in the department using scrap metal, local wood and plastics. To melt the scrap metal, the foundry has improvised a pit-furnace.
>
> The locally manufactured spares have proved to be of high quality. It is estimated that Kiltex . . . saves 1.5 million shillings annually in . . . foreign exchange . . . [and] earns about 25,000 shillings annually from the sale of spare parts to other TEXCO group companies.[52]

Pakistani and Tanzanian engineers and technicians were also responsible for the design of an oil facility that permitted the firm to reduce by 7 percent its annual consumption of oil.[53]

Finally, the engineering department developed and built a new machine to process waste cotton into surgical cotton and adapted existing equipment to make use of locally produced cornstarch.

In seeking to explain the process of technological apprenticeship launched within this firm since 1974, a team of researchers, on the basis of interviews with the KILTEX senior staff, concluded that

> the major stimulus has been the chronic foreign exchange problem. . . . This has severely restricted the exchange allocation for industrial raw materials and spare parts. . . . Kiltex . . . has come to regard the shortage of foreign exchange as a blessing in disguise: a stimulus to the manufacture of spare parts locally. Consequently, the firm has established an incentive package to all its workers. The incentives include: 1) rapid promotion, 2) double increment, 3) special monetary tokens in cases where innovations are put to immediate use and 4) training.[54]

Not only has KILTEX embarked upon a process of technological apprenticeship, but the decision to institute incentive policies points to the explicitly conscious fashion in which this process is being reinforced.

CONCLUSION

This chapter has sought to examine that set of factors that appears to have promoted or retarded an effective transfer of textile technology in Africa. An effective technology transfer, it was suggested, is essentially a process of building indigenous technological capabilities through a conscious effort to engage in learning-by-doing within the firm. The process of technological apprenticeship that this implies appears to be triggered when persons in key decision-making positions within an enterprise are responsive to their domestic environment, are sufficiently aware of technical possibilities, and are placed in a situation in which there are pressures to reduce production costs.

The African context has been a singularly unpropitious environment in which to expect a process of technological apprenticeship to emerge. As the Ivorian and Nigerian cases illustrate, most textile firms were initiated and managed by foreign owners or partners. High levels of tariff protection demanded by foreign investors, subsidized capital costs inadvertently offered by local governments, and low levels of domestic competition combined to shape initial choices of technique in directions that made continued reliance on expatriate personnel and imported inputs more likely. Price fixing in the local market and a competitive structure that favored product differentiation of an import reproductive nature reduced the incentive to embark upon the riskier process of technological innovation. Where there were few sanctions imposed as a result of the relatively high costs of production in these firms, and especially in cases in which state intervention further reduced the discipline of the market, firms were notably disinclined to embark upon a process of technological apprenticeship.

Even within this context, as the Kenyan and Tanzanian cases demonstrate, conditions could emerge to stimulate the process of learning-by-doing. Given Kenyan political realities, Asian textile manufacturers were obliged to adapt to a competitive structure shaped by the close ties established between a rising African bourgeoisie and foreign multinational firms. The choice of product and technique flowed from the manufacturers' unique situation within the Kenyan political economy and contributed to the ease with which they mastered the imported technology with which they launched their firms. In the Nigerian case, on the other hand, where competition on the basis of product differentiation became a key feature of the growth of the textile industry, quite another dynamic was encountered – one less likely to induce technological apprenticeship and more likely to lead to continued technological imports as a means of altering product mixes or increasing output.

While there is considerable evidence to support the conclusion that MNC textile subsidiaries have been least likely to embark on a process of technological apprenticeship in Africa, recent analyses have also pointed to a tendency among state-owned firms to choose techniques of production that are more capital intensive than those of other firms in their sector, to be more dependent upon imported inputs and expatriate technical assistance, and thus to be less likely to pursue a policy of technological apprenticeship than locally owned private firms.[55] In large part, however, these failures are due to the access to credit that state firms possess and the likelihood that restrictions on imports or on the use of expatriate personnel imposed on private firms will be waived in the case of state ventures, which by definition come to assume priority status. Just as state policies can shape the decision to replace the creation of indigenous technological capabilities by technology imports, so too, as the KILTEX case illustrates, might alternative policies influence a movement toward technological mastery, once appropriate technical skills have been acquired by those likely to respond to such policy changes.

As the Kenyan and Tanzanian cases, moreover, illustrate, of pivotal importance is the extent to which changes in competitive structures or in state policies are translated into a need to reduce costs of production, whether these be costs of foreign exchange, energy, or expatriate salaries. It is this need to reduce the costs of production that thus provides a powerful stimulus to technological mastery as it inevitably requires the assimilation of imported know-how by local personnel and the replacement of imports by domestic inputs – the latter inducing the kinds of product and/or process modifications that constitute steps in a process of technological apprenticeship.

NOTES

1. G. S. Gibb, *The Saco-Lowell Shops, Textile Machinery Building in New England, 1813-1849* (Cambridge, Mass: Harvard University Press, 1950); D. Jeremy, "Damming the Flood: British Government Efforts to Check the Outflow of Technicians and Machinery 1780-1843," *Business History Review* 51(1) (1977), 1-34.; D. Landes, *The Unbound Prometheus: Technological Change and Industrial Development in Western Europe from 1750 to the Present* (Cambridge: Cambridge University Press, 1969).

2. Takeo Izumi, "Transformation and Development of Technology in the Japanese Cotton Industry" monograph (Tokyo: United Nations University, 1980): p. 1.

3. Yung W. Rhee and Larry E. Westphal, "A Micro Econometric Investigation of Choice of Technology," *Journal of Development Economics* 4 (1977): 205-38.

4. Amitai Rath, "Technological Change, Transfer and Diffusion in the Textile Industry in India," in International Development Research Centre (IDRC), *Absorption and Diffusion of Imported Technology*, proceedings of a workshop held in Singapore, 26-30 January 1981 (Ottawa: IDRC, 1983), Doc. No. IDRC 171e: 49-60; Ashok V. Desai, "Technology and Market Structure under Government Regulation: A Case Study of the Indian Textile Industry," in IDRC, *Absorption and Diffusion*: 61-72.

5. G. Boon. *Technology Transfer in Fibres, Textile and Apparel* (Rockville, Md.: Sijthoff & Nourdhoff, 1981).

6. K. Arrow, "The Economic Implications of Learning by Doing," *Review of Economic Studies* 19 (1962): pp. 155-72.

7. M. Bell and K. Hoffman, *Industrial Development with Imported Technology: A Strategic Perspective on Policy* (Brighton: University of Sussex, Science Policy Research Unit, 1981), manuscript, p. 127.

8. P. David, "The 'Horndal Effect' in Lowell, 1834-56: A Shortrun Learning Curve for Integrated Cotton Textile Mills," in *Technical Choice, Innovation and Economic Growth*, ed. P. A. David (Cambridge: Cambridge University Press, 1975), pp. 174-91.

9. S. Hollander, *The Sources of Increased Efficiency: A Study of DuPont Rayon Plants* (Cambridge, Mass.: MIT Press, 1965).

10. M. A. Bell, D. Scott-Kemmis, and W. Satyarakwit, *Learning and Technical Change in the Development of Manufacturing Industry: A Case Study of Permanently Infant Enterprise* (Brighton: University of Sussex, Science Policy Research Unit, 1980).

11. Q. Ahmed, "Accumulation of Technological Capability and Assimilation of Imported Industrial Production Systems in the Cotton Textile Industry in Bangladesh," in IDRC, *Absorption and Diffusion*: 21-27; Bell and Hoffman, *Industrial Development*; H. Pack, "Productivity and Technical Choice: Applications to the Textile Industry," paper presented at the meeting of the American Economic Association, New York, December 1982.

12. Bell et al. *Learning and Technical Change*, p. 45.

13. C. J. Dahlman and L. E. Westphal, *Technological Effort in Industrial Development – An Interpretive Survey of Recent Research* (Washington, D.C.: Work Bank, 1981); J. Katz, M. Gurkowski, M. Rodrigues, and G. Goity, *Productivity, Technology and Domestic Efforts in Research and Development*, IDB/ECLA Research Programmes on Scientific and Technological Development in Latin America, Working Paper no. 13, July 1978; P. Maxwell, "Technical and Organization Changes in Steel Plants: An Argentine and a Brazilian Case," paper presented at a Simposio de Analysis Organizacional, Buenos Aires, 16-18 October 1980.

14. C. Cooper, "Science Policy and Technological Change in Underdeveloped Economies," *World Development* 2 (March, 1974): 55-64; Norman Girvan, "The Approach to Technology Policy Studies," *Social and Economic Studies* 28 (March, 1979): 1-53.

15. N. Rosenberg, *Perspectives on Technology* (Cambridge: Cambridge University Press, 1976), especially pp. 108-25, "The Direction of Technological Change: Inducement Mechanisms and Focusing Devices"; F. R. Sagasti, "Underdevelopment, Science and Technology: The Point of View of the Underdeveloped Countries," *Science Studies* 3 (1973): 47-59; P. Maxwell and M. Teubal, "Capacity-Stretching Technical Change: Some Empirical Theoretical Aspects," IDB/ECLA Research Programme on Scientific and Technological Development in Latin America, Working Paper no. 36, November 1980, pp. 13-48.

16. F. Stewart, *Technology and Underdevelopment* (London: Macmillan, 1977).

17. Girvan, "The Approach to Policy Studies," p. 14.

18. Bell and Hoffman, *Industrial Development*, p. 112.

19. Bell et al. "Learning and Technical Change."

20. Ediafric, *L'Industrie africaine en 1982*, 8^{e} edition, tome 2, "Côte d'Ivoire" (Paris: Ediafric, 1982): p. 7.

21. Ibid., p. 6.

22. Ibid., p. 7.

23. During the last years of colonialism, the Ivorian state became the locus of collaboration between an Ivorian planter class, whose basis for capital accumulation lay in the production of export crops, and French capital, whose interests were then primarily centered in commerce. By independence, this symbiotic relationship between foreign and Ivorian capital had been concretized in a division of spheres of capital accumulation according to which the agricultural sector was reserved to Ivorian capital, private and later state, while foreign capital was encouraged to expand into processing and secondary manufacturing. B. Campbell, "Neocolonialism, Economic Dependence and Political Change: Cotton Textile Production in the Ivory Coast," *Review of African Political Economy* 2 (1976): 36-53; J. Masini, M. Ikanicoff, C. Jedlicki, M. Lazarotti, *Les Multinationales et le développement: Trois Entreprises en Côte D'Ivoire* (Paris: PUF, 1979); L. K. Mytelka, "Direct Foreign Investment and Technological Choice in the Ivorian Textile and Wood Industries," *Vierteljahresberichte*, 83 (March 1981): 61-80; L. K. Mytelka, "The Limits of Export-Led Development: Ivory Coast's Experience with Manufacturers" in *The Antinomies of Interdependence*, ed. J. Ruggie (New York: Columbia University Press, 1983):

pp. 239-70; M. L. Mazoyer, "Développement de la production et transformation Agricole Marchande: D'une formation agraire en Côte d'Ivoire," in *L'Agriculture africaine et le capitalisme* ed. S. Amin (Paris: Ed. Anthropos, 1976): pp. 143-66; M. L. Mazoyer, "Côte-d'Ivoire," in *Pauvreté et inégalities rurales en Afrique de l'Ouest francophone*, ed. R. Dumont, C. Reboul, and M. L. Mazoyer (Geneva: BIT, 1981): pp. 50-73; C. de Miras, "L'Entrepreneur Ivoirien ou une bourgeoisie privée de son état," in *Etat et bourgeoisie en Côte-d'Ivoire*, ed. Y. Faure and J. F. Médard (Paris: KARTHALA/ORSTOM, 1981): pp. 181-230; J. M. Gastellu and S. Affou Yapi, "Un Mythe à Décomposer: La bourgeoisie de Planteurs," in *Etat et bourgeoisie en Côte-d'Ivoire*, ed. Y. A. Faure and J. F. Médard (Paris: KARTHALA/ORSTOM, 1983): pp. 149-80.

24. Mytelka, "Direct Foreign Investment."

25. J. Chevassu and A. Valette, *Donnés statistiques sur l'industrie de la Côte d'Ivoire* (Abidjan: ORSTOM, 1975); International Development Research Centre, *Absorption and Diffusion of Imported Technology*, proceedings of workshop, Singapore, 26-30 January 1981, doc. no. IDRC. 171e (Ottawa: IDRC, 1983).

26. J. Lecaillor and D. Germidis, *Inégalité des revenus et développement économique* (Paris: Presses Universitaires de France, 1977): p. 45.

27. Chevassu and Valette, *Donnés statistiques*, pp. 3, 5; Chevassu and Valette, *Les revenus distribués, parles activités industrielles en Côte D'Ivoire* (Abidjan: ORSTOM, 1975).

28. Michael Cohen, *Urban Policy and Political Conflict in Africa: A Study of the Ivory Coast* (Chicago: University of Chicago Press, 1974).

29. *La Côte d'Ivoire en chiffre* (Dakar: Sté Africaine d'Editions: 1976, 1978-79).

30. International Monetary Fund, *Balance of Payments Yearbook* (Washington, D.C.: IMF, various years).

31. Mytelka, "Direct Foreign Investment."

32. Ibid., pp. 75-76.

33. Mytelka, "The Limits of Export-Led Development."

34. *Bulletin de l'Afrique Noire*, no. 1158 (November 12, 1982).

35. *Bulletin de l'Afrique Noire*, no. 1121 (January 1, 1982).

36. Ibid., no. 1124 (January 4, 1982).

37. "Interview with M. Yves Philippe, Secretary General of Establissements Gonfreville," *Afrique Industrie*, March 15, 1980, pp. 64-67.

38. S. Langdon, "Industry and Capitalism in Kenya: Contributions to a Debate," paper prepared for a conference on "The African Bourgeoisie: The Development of Capitalism in Nigeria, Kenya and the Ivory Coast, Dakar, December 2-4, 1980.

39. Ibid., pp. 28-29. See also R. Kaplinsky, "Capitalist Accumulation in the Periphery – the Kenyan Case Re-examined," *Review of African Political Economy* No. 17 (Jan./Apr. 1980): 83-105; and N. Swainson, *The Development of Corporate Capitalism in Kenya, 1918-1977* (Berkeley: University of California Press, 1980). The potential of indigenous capitalism in Kenya is hotly debated.

40. L. K. Mytelka, "Licensing and Technology Dependence in the Andean Group," *World Development* 6 (1978): 447-59.

41. S. W. Langdon, "Indigenous Technological Capacity in Africa: The Case of Textiles and Wood Products in Kenya," paper prepared for the International Workshop on Facilitating Indigenous Technological Capacity, Centre of African Studies, Edinburgh University, May 1982, p. 10.

42. Ibid., pp. 15-16.

43. Ibid., p. 12.

44. S. W. Langdon, *Multinational Corporations in the Political Economy of Kenya* (London: Macmillan, 1981), especially pp. 48-64, "Subsidiaries, Product Reproduction and Taste Transfer," and pp. 65-97, "Multinational Corporations and Kenya Industry: Two Case Studies."

45. A. Amsden, "The Competitive Process Under Capitalism: Machine Tool Building in a Third World Context," Department of Economics, Barnard College, Columbia University, unpublished paper.

46. C. N. S. Nambudiri, O. Iyanda, and D. M. Akinnusi, "Third World Country Firms in Nigeria," in *Multinationals from Developing Countries*, ed. K. Kumar and M. G. McLeod (Lexington, Mass.: Lexington Books, 1981): p. 147.

47. On the basis of research conducted in Nigeria during the mid-1960s, Gerald K. Helleiner noted that "Nigerian manufacturing technology, like that of other underdeveloped areas, seems to have been imported from abroad without major modifications," and he added, ". . . given the existing degree of overt urban unemployment the degree of capital-intensity of the new plants is distressingly high." See Gerald K. Helleiner, *Peasant Agriculture, Government, and Economic Growth in Nigeria* (Homewood, Ill.: Richard D. Irwin, 1966): p. 33.

48. Maxwell, *Technical and Organization Changes in Steel Plants.*

49. Rosenberg, *Perspectives on Technology.*

50. C. E. Barker et al., *Industrial Production and Transfer of Technology in Tanzania* (Dar-es-Salaam: Institute for Development Studies, 1978).

51. S. Rugumamu et al., "Issues in Technology Assimilation: The Case of Kilimanjaro Textiles of Arusha, Tanzania," mimeo (Arusha: IDRC/IDS Workshop, April 1982): p. 3.

52. Ibid., p. 5.

53. Ibid.

54. Ibid., p. 6.

55. A. Coulson, "The State and Industrialization in Tanzania," in *Industry and Accumulation in Africa*, ed. M. Fransman (London: Heinemann, 1982): 60-76.

REFERENCES

Q. Ahmed. "Accumulation of Technological Capability and Assimilation of Imported Industrial Production Systems in the Cotton Textile Industry in Bangladesh." In IDRC *Absorption and Diffusion*, pp. 21-27.

M. Amsalem. *Technology Choice in Developing Countries: The Textile and Pulp and Paper Industries*. Cambridge, Mass.: MIT Press, 1983.

A. Amsden. "The Competitive Process Under Capitalism: Machine Tool Building in a Third World Context." Barnard College, Columbia University, Department of Economics. Unpublished, n.d.

K. Arrow. "The Economic Implications of Learning by Doing." *Review of Economic Studies* 29 (1962): pp. 155-72.

C. E. Barker, M. R. Bhagavan, P. M. von Mitschke-Collande, and D. V. Wield. *Industrial Production and Transfer of Technology in Tanzania*. Dar-es-Salaam: Institute for Development Studies, 1978.

M. Bell and K. Hoffman. *Industrial Development with Imported Technology: A Strategic Perspective on Policy*. Brighton: University of Sussex, Science Policy Research Unit, 1981. Manuscript.

M. A. Bell, D. Scott-Kemmis, and W. Satyarakwit. *Learning and Technical Change in the Development of Manufacturing Industry: A Case Study of Permanently Infant Enterprise*. Brighton: University of Sussex, Science Policy Research Unit, 1980.

M. Bienefeld. "Evaluating Tanzanian Industrial Development." In *Industry and Accumulation in Africa*, edited by Martin Fransman. London: Heinemann, 1982.

T. J. Biersteker. *Distortion or Development? Contending Perspectives on the Multinational Corporation*. Cambridge, Mass.: MIT Press, 1978.

G. Boon. *Technology Transfer in Fibres, Textile and Apparel*. Rockville, Md.: Sijthoff & Nourdhoff, 1981.

Bulletin de l'Afrique Noire. Various issues.

B. Campbell. "Ivory Coast." In *West African States: Failure and Promise*, edited by J. Dunn. Cambridge: Cambridge University Press, 1978.

____. "Neocolonialism, Economic Dependence and Political Change: Cotton Textile Production in the Ivory Coast." *Review of African Political Economy* 2 (1976): pp. 36-53.

J. Chevassu and A. Valette. *Donnés statistiques sur l'industrie de la Côte d'Ivoire*. Abidjan: ORSTOM, 1975.

____. *Les Revenus distribués par les activitiés industrielles en Côte d'Ivoire*. Abidjan: ORSTOM, 1975.

M. Cohen. *Urban Policy and Political Conflict in Africa: A Study of the Ivory Coast*. Chicago: University of Chicago Press, 1974.

C. Cooper. "Science Policy and Technological Change in Underdeveloped Economies." *World Development* 2 (March 1974): pp. 55-64.

Côte d'Ivoire, République de, Ministère de l' Economie, des Finances et du Plan. *Memorandum textile pour la Communauté Economique Européenne*. Abidjan: December, 1978.

La Côte d'Ivoire en chiffre. Dakar: Sté Africaine d'Editions, 1976, 1978-79.

A. Coulson. "The State and Industrialization in Tanzania." In *Industry and Accumulation in Africa*, edited by M. Fransman. London: Heinemann, 1982, pp. 60-76.

C. Dahlman and L. E. Westphal. *Technological Effort in Industrial Development – An Interpretive Survey of Recent Research*. Washington, D.C.: World Bank, 1981.

P. David. "The 'Horndal Effect' in Lowell, 1834-56: A Shortrun Learning Curve for Integrated Cotton Textile Mills." In *Technical Choice, Innovation and Economic Growth*, edited by P. A. David. Cambridge: Cambridge University Press, 1975, pp. 174-91.

A. V. Desai. "Technology and Market Structure Under Government Regulation: A Case Study of the Indian Textile Industry." In IDRC, *Absorption and Diffusion*, pp. 61-72.

Ediafric. *L'Industrie africaine en 1982*, 8e edition, tome 2, "Côte d'Ivoire." Paris: Ediafric, 1982.

T. Forrst. "Recent Developments in Nigerian Industrialization." In *Industry and Accumulation in Africa*, edited by M. Fransman. London: Heinemann, 1982, pp. 324-44.

J. M. Gastellu and S. Affou Yapi. "Un Mythe à décomposer: La bourgeoisie de Planteurs." In *Etat et bourgeoisie en Côte-d'Ivoire*, edited by Y. A. Faure and J. F. Médard. Paris: KARTHALA/ORSTOM, 1983, pp. 149-80.

G. S. Gibb. *The Saco-Lowell Shops, Textile Machinery Building in New England, 1813-1849*. Cambridge, Mass.: Harvard University Press, 1950.

N. Girvan. "The Approach to Technology Policy Studies." *Social and Economic Studies* 28 (March, 1979): pp. 1-53.

G. K. Helleiner. *Peasant Agriculture, Government, and Economic Growth in Nigeria*. Homewood, Ill.: Richard D. Irwin, 1966.

S. Hollander. *The Sources of Increased Efficiency: A Study of DuPont Rayon Plants*. Cambridge, Mass.: MIT Press, 1965.

A. Hoogvelt. "Indigenization and Foreign Capital: Industrialization in Nigeria." *Review of African Political Economy* 14 (January-April 1979): pp. 56-68.

International Development Research Centre, *Absorption and Diffusion of Imported Technology*. Proceedings of a workshop, Singapore, 26-30 January 1981. Doc. no. IDRC. 171e. Ottawa: IDRC, 1983.

International Monetary Fund. *Balance of Payments Yearbook*. Washington, D.C.: IMF, various years.

"Interview with M. Yves Philippe, Secretary General of Etablissements Gonfreville." *Afrique Industrie*, March 15, 1980, pp. 64-67.

T. Izumi. "Transformation and Development of Technology in the Japanese Cotton Industry." monograph. Tokyo: United Nations University, 1980.

D. Jeremy. "Damming the Flood: British Government Efforts to Check the Outflow of Technicians and Machinery 1780-1843." *Business History Review* (1977), 51(1), pp. 1-34.

R. Kaplinsky. "Capitalist Accumulation in the Periphery – The Kenyan Case Re-examined." *Review of African Political Economy*, No. 17 (January-April 1980): pp. 83-105.

J. Katz, M. Gutkowski, M. Rodrigues, and G. Goity. *Productivity, Technology and Domestic Efforts in Research and Development*. IDB/ECLA Research Programmes on Scientific and Technological Development in Latin America, Working Paper no. 13, July 1978.

M. Kuuya. "Import Substitution as an Industrial Strategy: The Tanzanian Case." In *Industrialization and Income Distribution in Africa*, edited by J. F. Rweyemamu. Dakar: Codesria, 1980, pp. 69-91.

D. Landes. *The Unbound Prometheus: Technological Change and Industrial Development in Western Europe from 1750 to the Present*. Cambridge: Cambridge University Press, 1969.

S. Langdon. "Indigenous Technological Capacity in Africa: The Case of Textiles and Wood Products in Kenya." Paper prepared for the International Workshop on Facilitating Indigenous Technological Capacity, Centre of African Studies, Edinburgh University, May 1982.

———. "Industry and Capitalism in Kenya: Contributions to a Debate." Paper prepared for a conference on "The African Bourgeoisie: The Development of Capitalism in Nigeria, Kenya and the Ivory Coast," Dakar, December 2-4, 1980.

———. *Multinational Corporations in the Political Economy of Kenya*. London: Macmillan, 1981.

G. Larson. "La Côte d'Ivoire: 1906-70, Croissance et diversification sans Africanisation." In *L'Afrique de l'indépendance politique à l'indépendance économique*, edited by J. D. Esseks. Paris: Maspero and Presses Universitaires de Grenoble, 1975, pp. 201-38.

J. Lecaillor and D. Germidis. *Inégalité des revenues et développement économique*. Paris: Presses Universitaires de France, 1977.

J. Masini, M. Ikanicoff, C. Jedlicki, and M. Lanzarotti. *Les Multinationales et le développement: Trois entreprises en Côte d'Ivoire*. Paris: PUF, 1979.

P. Maxwell. "Technical and Organization Changes in Steel Plants: An Argentine and a Brazilian Case." Paper presented at a Simposio de Analysis Organizacional, Buenos Aires, 16-18 October 1980.

P. Maxwell and M. Teubal. "Capacity-Stretching Technical Change: Some Empirical and Theoretical Aspects." IDB/ECLA Research Programme on Scientific and Technological Development in Latin America, Working Paper no. 36, November 1980.

M. L. Mazoyer. "Côte-d'Ivoire." In *Pauvreté et inégalités rurales en Afrique de l'Ouest francophone*, edited by René Dumont, Claude Reboul, and Marcel Mazoyer. Geneva: BIT, 1981, pp. 50-73.

———. "Développement de la production et transformation agricole marchande: D'une formation agraire en Côte d'Ivoire." In *L'Agriculture africaine et le capitalisme*, edited by S. Amin. Paris: Ed. Anthropos, 1976, pp. 143-66.

C. de Miras. "L'Entrepreneur Ivoirien ou une bourgeoisie Privée de son état." In *Etat et bourgeoisie en Côte-d'Ivoire*, edited by Y. Faure and J. F. Médard. Paris: KARTHALA/ORSTOM, 1981, pp. 181-230.

L. K. Mytelka. "Direct Foreign Investment and Technological Choice in the Ivorian Textile and Wood Industries," *Vierteljahresberichte* 83 (March 1981): pp. 61-80.

———. "Licensing and Technology Dependence in the Andean Group," *World Development* 6 (1978): pp. 447-59.

———. "The Limits of Export-Led Development: Ivory Coast's Experience with Manufacturers." In *The Antinomies of Interdependence*, edited by J. Ruggie. New York: Columbia University Press, 1983, pp. 239-70.

C. N. S. Nambudiri, O. Iyanda, and D. M. Akinnusi. "Third World Country Firms in Nigeria." In *Multinationals from Developing Countries*, edited by K. Kumar and M. G. McLeod. Lexington, Mass.: Lexington Books, 1981, pp. 145-53.

H. Pack. "Productivity and Technical Choice: Applications to the Textile Industry." Paper presented at the meeting of the American Economic Association, New York, December 1982.

W. H. Park. "Absorption and Diffusion of Imported Technology: A Case Study in the Republic of Korea." In *Absorption and Diffusion*, 79-89.

J. Pickett and R. Robson. "Production of Cotton (Cloth) with Special Reference to African Conditions." In UNIDO, *Appropriate Industrial Technology for Textiles*. Monographs on Appropriate Industrial Technology, no. 6. New York: United Nations, 1979, pp. 19-36.

A. Rath. "Technological Change, Transfer and Diffusion in the Textile Industry in India." In IDRC, *Absorption and Diffusion*, 49-60.

Y. W. Rhee and L. E. Westphal. "A Micro Econometric Investigation of Choice of Technology." *Journal of Development Economics* 4 (1977): pp. 205-38.

N. Rosenberg. *Perspectives on Technology*. Cambridge: Cambridge University Press, 1976.

S. Rugumamu, C. Mupimpila, T. Olembo, and S. Sanyila. "Issues in Technology Assimilation: The Case of Kilimanjaro Textiles of Arusha, Tanzania." Arusha: IDRC/IDS Workshop, April 1982. Mimeograph.

F. Sagasti. "The Itintec System for Industrial Technology Policy in Peru." *World Development* 3 (1975): pp. 827-37.

——. "Underdevelopment, Science and Technology: The Point of View of the Underdeveloped Countries." *Science Studies* 3 (1973): pp. 47-59.

F. Stewart. *Technology and Underdevelopment*. London: Macmillan, 1977.

N. Swainson. *The Development of Corporate Capitalism in Kenya, 1918-1977*. Berkeley: University of California Press, 1980.

United Nations Industrial Organization. "Recent Industrial Development in Africa." In *Industry and Accumulation in Africa*, edited by M. Fransman. London: Heinemann, 1982, pp. 386-415.

G. Winston. "The Appeal of Inappropriate Technologies: Self-Inflicted Wages, Ethnic Pride and Corruption." *World Development* 7 (1979): pp. 835-45.

5

Domestic Technological Innovations and Dynamic Comparative Advantages: Further Reflections on a Comparative Care-Study Program

Jorge M. Katz

INTRODUCTION

The development of indigenous technological capabilities in developing countries is scarcely understood. Our knowledge is even poorer in such related fields as dynamic shifts in comparative advantages across industries and countries, which require a thorough understanding of the process through which industries or countries accumulate technical knowledge and skills over time.

Conventional theory is not particularly useful. On the one hand, it has never seriously entertained the idea that developing countries might eventually be anything but passive recipients of foreign blueprints taken from the "freely available international shelf" (whatever that metaphor might mean, given the realities of the international technological market). On the other hand, it has never considered the fact that technical knowledge is a very special input indeed.[1] As a consequence, conventional theory has never quite understood that firms or countries have concomitantly to generate technical knowledge in order to use technical knowledge, and that different firms or countries do so at different rates and with varying success.

A previous version of this paper was presented by the author at the American Economic Association meeting held in New York in December 1982. Most helpful comments by G. Ranis and M. Bell are acknowledged.

Yet economic history and casual observation tell us a different story. Firms, industries, and countries do from time to time lose their technological leadership in specific fields to rapidly learning followers. These countries, and subsequently others, catch up on the basis of either lower production costs or better and improved technological design. Obviously the development of indigenous technological capabilities of various types must necessarily be assumed to underlie this phenomenon.

It has taken a long time – from Leontief's paradox to the writings of authors such as M. Peck or H. Patrick[2] – for the profession to apply such a "catch-up" model to the contemporary Japanese experience. The profession has been slow to recognize that countries such as Brazil, South Korea, Spain, and Argentina are now exporting sophisticated industrial goods with a significant amount of domestically generated technical knowledge embodied in them in the form of licenses and engineering services, turnkey manufacturing plants, and infrastructural works, construction services, and so on.[3] The profession is even less prepared to believe that some of these newly industrializing countries (NICs) are also acting as capital exporters, erecting foreign subsidiaries or entering into partnership in joint-venture agreements, both in their neighboring surroundings and in more distant, relatively poorer developing countries.

I shall argue here that these recent trends constitute prima facie evidence that major structural changes have been taking place for some time in the domestic technological capabilities of these countries, changes that conventional scholarship has failed to capture and understand. Moreover, the development of domestic technological capabilities is part of a more complex social phenomenon in which the organization of work and the social division of labor undergo qualitative changes.[4] The overall production fabric of these sectors suffers structural changes with the emergence of new markets, new institutions, new informational channels, and new forms of specialization. In other words, the development of domestic technological capabilities should be regarded as part of a more general maturing process about which we understand precious little. In spite of its attractiveness, I shall not, however, deal with this more general proposition in this chapter.

Instead, I shall concentrate on the strictly technological side of the maturing process. New technical and engineering activities such as quality control, preventive maintenance, normalization and standardization of inputs, parts, components, time and motion studies, and production planning and organization, have gradually appeared and become routine in firms. Concomitantly, a different attitude toward engineering rules and norms gradually permeated through society at large.* Many of these new engineering activities tend to be localized in specialized technical departments of medium-size and large firms. Such departments generate incremental units of technical knowledge in the areas of product design, process engineering, and production planning and organization. They adapt foreign technology to the local environment and gradually build up a stock of proprietary technology and know-how highly specific to the firm. It is here that domestic technological capabilities actually appear and develop.

I examine below the technological and economic performance of more than 50 industrial firms over several decades of their operation in the 6 largest Latin American countries. My purpose throughout the inquiry is not to attain statistical representativeness. Pin factories are not known in the profession for their representativeness, but they nonetheless have been the source of one of the most powerful insights in the history of economics. Instead, our purpose is to produce some initial building blocks for a more ambitious future construction. I found justification for this method in the vast amount of literature that provides the basis of conventional wisdom but is seldom grounded on empirical observation.

In the large majority of these case studies I enjoyed ample collaboration from the firms over a long period of time. The inquiry is based on a vast amount of otherwise unavailable information concerning the determinants and consequences of entrepreneurial behavior in general and of technological decisions in particular, that is, change in the package of technical knowledge and information

*It is worth noting that I am by no means arguing here that the process of structural change is evenly distributed in an intersectoral, interregional, or interpersonal sense. Obviously what is true of the so-called *triańgulo mineiro* (the area located inbetween Rio de Janeiro, Saô Paulo and Belo Horizonte) is certainly not true in Brazil's northeastern region. A similar pattern of uneven development can be uncovered in the early years of the industrial revolution in the United Kingdom or in the United States. Recognizing the unbalanced nature of the distribution process should not, however, impede the exploration of the underlying structural trends.

used by any given enterprise in the areas of product design, process engineering, and production organization.

The study covers a wide range of underlying conditions in the size, nationality, type of industry, and patterns of industrial organization. In other words, I examine small family enterprises as well as large incorporated firms, locally owned as well as foreign-owned companies, firms producing steel, cement, and petrochemicals, that is, continuous-flow commodities as well as plants producing agricultural machinery and machine tools where production runs tend to be small and where the organization of production tends to be discontinuous.

The second section of the chapter characterizes the central actor of our drama, manufacturing firms. Because received theory is of little help in this field, I constructed a simple typology to reflect observed differences in domestic technological capabilities between family enterprises, local subsidiaries of multinational corporations (MNCs), and public firms, as well as between firms in developing countries and their counterparts in more developed industrial societies.

The third section examines the micro and macro determinants of the domestic technological search efforts. The fourth explores the sequential nature of the development of domestic technological capabilities in two polar cases, first in a family-owned small-batch mechanical engineering firm, and second, in a continuous-flow process plant. The final section deals briefly with the relationship between the development of indigenous technological capabilities and infant industry protection policies of the state.

TYPOLOGICAL DIFFERENCES AND THE DEVELOPMENT OF DOMESTIC TECHNOLOGICAL CAPABILITIES

The kind of technological capabilities that emerge and develop in any given social setting strongly depends on the type of economic agents in such setting, the resource endowments they control, and the public policies by which they are affected over time. Consider first the role of the agents themselves. It seems reasonable to expect that a small, family mechanical engineering enterprise that produces small batches of different products in a discontinuous plant layout will call for the development of different indigenous technological capabilities than will a large, foreign-owned, continuous-flow plant that produces a single, quasi-homogeneous, and standardized commodity, such as steel or petrochemicals.

Thus the size and nature of the firm, its field of activity, type of production organization, degree of product standardization, and type of ownership all seem to be important determining factors in the development of indigenous technological capabilities. Over and beyond such factors, a vast array of policy-induced actions affect both the rate and the nature of the domestically performed technological search efforts. Such efforts will obviously underlie the development of indigenous technological capabilities of any sort. Let us begin by examining some of these factors, indicating – wherever possible – their likely effect upon the development of indigenous technological capabilities.

Type of Firm

It seems appropriate to begin with manufacturing firms. It is scarcely surprising that conventional theory is not a great help in this territory. The neoclassical textbook model of the firm is much too simple for the task we have in front of us. As R. Nelson puts it:

> Firms are the key productive actors, transforming inputs into outputs according to a production function. The production function, which defines the maximum output achievable with any given quantity of inputs, is determined by the state of technological knowledge. Technological knowledge is assumed to be public or at least this is implicit in models based on an industry or an economy-wide production function. Firms choose a point on their production function to maximize profits, given product demand and factor supply conditions. Generally these markets are assumed to be perfectly competitive so that a firm treats prices as parameters. . . . Over time, output grows as inputs increase and firms move along their production functions, and as technology advances. . . . There clearly are some strong presumptions here. The view of firms and markets is very stylized – not much room for incompetent management, labor-management strife, or oligopolistic rivalry. Technological advance, while acknowledged as a central feature of growth, is treated in a very simple way, and the Schumpeterian proposition that technological advance (via entrepreneurial innovation) and competitive equilibrium cannot co-exist is ignored. . . .

> (F)rom the neoclassical perspective, there are few interesting empirical questions that can be explored or resolved by studying particular firms or by considering differences among individual firms in similar market conditions.[5]

Given such a state of affairs, I start by constructing a simple typology of manufacturing firms that could become useful in the exploration of behavioral differences in the development of domestic technological capabilities.

Four types of firm need to be singled out for this purpose: family enterprises, domestic subsidiaries of large MNCs, public companies, and large domestically owned incorporated firms. It is important to understand right from the start that these types of firm differ quite significantly in aspects such as production organization, access to technical information and factor markets, and long-term objectives and constraints. Such differences seem to me to be essential aspects of reality that cannot be sacrificed in order to attain a higher level of generalization. Both the rate and the nature of the long-term expansion path of any given industry are going to be affected by the types of firm that belong in it, and it seems wrong to suppress their differences a priori. Let us go into them in some more detail.

Family Enterprises

Family firms are characterized by the extremely narrow circle of reference to which they are attached, as well as by the high incidence of a number of extra-economic factors underlying their daily operation. A high proportion of self-designed and self-constructed pieces of capital equipment are usually found at the shop level, where much more attention is normally paid to mechanical engineering skills than to production planning and organizational activities.

A high propensity for self-financing is normally part of the package. To some extent such behavior originates in the more difficult access these companies have to capital markets, but one should not readily discard the role of tight family links underlying this pattern. The use of nonmarket criteria in the screening and hiring of technical personnel, as well as in capital equipment procurement, is also standard practice among family firms.[6]

Perhaps a somewhat more sophisticated description of this type of enterprise would take into account the existence of at least two major subgroups. There are, on the one hand, those closely run family firms in which the original owner still maintains strong control over everyday decisions. On the other hand, there are also numerous family enterprises that have already gone through a major generational shake-up that has brought to their management boards a younger generation of "engineers, lawyers, and graduates in business administration"[7] who replace the older generation of self-made mechanics.

Whereas family enterprises in the latter category are undoubtedly characterized by greater professionalism, those in the former probably maintain more of the Schumpeterian "animal spirits" that were congenial to the older generation. In any case, one should probably expect different responses from them in aspects such as risk taking, the relative significance attached to technical and mechanical aspects vis-à-vis commercial and financial ones, interest in the domestic market vis-à-vis the international one, lobbying capacity in the domestic political arena, and so on.

The empirical evidence collected during the course of my research also points out another interesting topic as far as family enterprises are concerned. This time the finding refers to intercountry differences within the broad set of family firms examined during the course of the enquiry. Probably in response to a bigger domestic market and also as a consequence of a more stable and favorable overall macropolicy framework, Brazilian family enterprises show a greater propensity than their counterparts in, say, Argentina or Colombia, to 1) move toward plant scales closer to international standards, 2) carry on larger domestic R & D programs, and 3) export a larger proportion of their current output, making particular inroads in markets in developed countries.[8]

Local Subsidiaries of MNCs

Domestic subsidiaries of large MNCs are different from the previous group in spite of the strong effort they normally make to mimic local production organization. There are first the start-up differences in technological, financial, organizational, and other aspects resulting from the fact that investment plans and engineering blueprints are normally prepared and decided by headquarters.

Sometimes a conscious effort is made by the parent's engineering and management staff to adapt these to the local environment, but such efforts are not always made, nor are they carried out in great detail. Thus external influences make domestic subsidiaries of MNCs special production agents when compared with local firms.

But obviously the story does not finish at the design and start-up period. At least two other sets of circumstances seem to be present in the years that follow. On the one hand, domestic subsidiaries of MNCs try to adapt themselves to the local environment and for this purpose develop domestic technological capabilities of their own in various fields. On the other hand, and because they normally are an appendix of the parent company, domestic subsidiaries do not need to develop engineering strength in each and every technical field the way local firms do. This seems to be particularly true in product design engineering, an area in which the availability of successive product generations filed in the headquarters technical library means that they need to make relatively smaller efforts in new product design, concentrating on the adaptation of the product to the local environment.

Given prevailing differences at the shop level between these firms and locally owned ones – in terms of production engineering, the make-buy ratio, and subcontracting practices (see below) – domestic subsidiaries of MNCs tend to develop relatively stronger domestic technological capabilities in process engineering and production planning and organization than in new product design.*

Finally, it is important to notice that even if it is true that domestic subsidiaries of MNCs enjoy an advantageous position as a consequence of their access to the pool of technical knowledge of the parent companies, it is also true that they pay tribute for this advantage in terms of less freedom of action and therefore less flexibility in their daily operation. In industries in which a quick response

---

*Compare, for example, the results obtained by A. Castaño et al. in their exploration of the technological behavior of Turri S.A., a locally owned machine tool firm, with those obtained by J. Berlinski in his study of Perkins S.A., a domestic subsidiary of a British firm, producing combustion engines in Argentina. See Castaño et al., "Etapas historicas"; J. Berlinski, "Innovaciones en el proceso y aprendizaje (El caso de una planta argentina de motores)," Monografía de Trabajo no. 59, Programa BID/CEPAL/CIID/PNUD de Investigaciones sobre Desarrollo Científico y Tecnológico en America Latina (Buenos Aires, September 1982), especially chap. 1.

constitutes a sine qua non for successful performance – such as, say, pharmaceuticals – this situation has permitted local entrepreneurs to compete successfully with them.[9]

Public Enterprises

A third, quite separate category is that of public enterprises. These are prominent in basic industries, such as steel, petroleum distillation, and petrochemicals. Many of these industries demand heavy initial capital investment, exhibit very long gestation and running-in periods (more than ten years are not unheard of in the case of a new steel mill), and are normally characterized by considerable physical imbalances among different operating sections and processes.

In various aspects public enterprises are very peculiar animals indeed. One such aspect is precisely the development of domestic technological capabilities of various sorts. On the one hand, and as far as choice of technique, plant design, procurement practices, and so on are concerned, public corporations have a strong tendency to go in for turnkey arrangements with international engineering contractors,[10] which, more often than not, try new processes and engineering designs at the expense of the public enterprises and involve them in risk taking.*

On the other hand, and given that their management is not normally subject to strict accountability and that profit incentives are frequently beyond their immediate interest, public firms usually spend much more time handling slack and out-of-equilibrium situations. Insofar as this frequently demands technological search

---

*Freeman's pioneering studies and some recent smaller contributions by Latin American writers on the operation of the international market for engineering services – design, procurement, plant construction, and the like – has still been hardly explored; the extent to which the contractual warranties normally underlying any specific deal are actually enforceable in court is highly questionable. When, where, and how the rights of the purchasing party are to be withheld is still a rather blurred field subject to deep controversy. See C. Freeman, "Chemical Process Plant: Innovation and the World Market," *National Institute Economic Review* no. 45 (August 1968); A. Aráoz, "Consulting and Engineering Design Organizations in Developing Countries," (December 1977), mimeograph; J. Zlotsky, *Seguimiento y control de estudios, proyectos y obras de ingeniería* (Buenos Aires: Editorial Ballesta 1976), chap. 3, p. 67.

efforts of various sorts, the public nature of ownership might very well be expected to affect the development of indigenous technological capabilities.

It is not, however, a matter of logical necessity for public enterprises to be examples of insufficient and/or incorrect development of domestic technological capabilities. Among the public firms studied during the course of this research program I had the opportunity to examine the case of USIMINAS, (Usima Siderúrgica de Minas Gerais), Brazil's largest steel mill, and also a very successful technological performer. This firm managed to evolve from an initial turnkey contract with a Japanese company to a position of technological self-sufficiency over the course of two decades. It is now exporting engineering services and acting as a training ground for technicians and engineers of various other publicly owned steel mills throughout Latin America,[11] giving proof that public ownership does not necessarily mean a wrong technological policy package.

Locally Owned Large Incorporated Firms

In various different aspects, such as access to technical information and local or foreign capital markets, and relative bargaining power vis-à-vis capital goods suppliers, large locally owned firms appear to be on much the same footing as domestic subsidiaries of MNCs. They have the right kind of international contacts for procurement purposes, and their managerial staff and technical personnel regularly attend international equipment fairs, technical seminars, and so forth.

As far as the development of domestic technological capabilities is concerned, large locally owned firms quite frequently have important spillover effects in that personnel from, for example, their engineering departments seek employment elsewhere in the manufacturing sector, carrying embodied skills and technical knowledge with them.

Having identified at least four different types of manufacturing firm and having pointed out some of their major similarities and differences in the development of indigenous technological capabilities, I now turn to a second typological aspect influencing such development. I refer to differences in production organization.

Production Organization

In a stylized fashion one could think of manufacturing production as being carried out with two rather different forms of production organization: continuous-flow manufacturing plants turning out large volumes of standard items, and small-batch, discontinuous plants, organized as a succession of shops or sections. Admittedly, this extreme characterization is solely for the sake of argument; a continuum of successive degrees of linearization probably constitutes a better description of reality than a simple bipolar model.* The simple dichotomy will, however, do for the purpose of exploring the major influence of alternative forms of production organization upon the development of domestic technological capabilities.

Consider first the case of a continuous-flow manufacturing plant. Manufacturing plants employing a continuous technology are product specific, that is, their layout is organized according to a sequence imposed by the various technical transformations that have to be carried out for the purpose of producing a given product. The sequence of technical transformation is always the same, and this is what determines the plant's layout. In manufacturing units of this sort the rate of output is usually rather large. Continuous-flow technologies are normally employed to produce massively commercialized products. Common features of a plant of this kind are:

1. The preproduction planning of the line is extremely detailed and complete. There is low ex-post flexibility in both product design and production process.

---

*In actual fact it would probably be better to think of metalworking plants as located along a continuum in which, at one extreme, would be the most elementary and fragmented shop layout and, at the other, a highly automated flexible manufacturing system, composed of 'isoles' of machining centers, robots, and automated equipment for transport and tool setting operations. In between would probably be various levels of plant linearization through successive implementation of the so-called group technology methodology of production planning and organization. See Organizacion Internacional del Trabajo (OIT), *Introducción al estudio del trabajo* (Ginebra: OIT, 1966); I. Miyata, "Economical Methods for Building a Flexible Manufacturing System," in *Metalworking Engineering and Marketing* vol. 5, no. 2 (Japan: March 1983): pp. 38-41; E. A. Arn, *Group Technology* (New York: Springer-Verlag, 1975); Castaño et al., "Etapas historicas."

2. Activities and technical transformation systematically follow one after the other along a direct route, thus minimizing delays and wasted time. The production cycle is minimized ex ante, as the line is balanced and activities have to be individually coordinated to the level of the micromovement.
3. Handling of raw materials and stocks of work in progress is also minimized. Inventories and storage spaces have to be balanced in conjunction with the overall production line.
4. The product tends to be highly normalized, and most of the equipment is of a rather specific nature, that is, specially designed to fulfill particular tasks or combinations of tasks.
5. There is relatively little on-the-job decision making, thus direct labor skills and supervisory requirements are relatively less important than in discontinuous production units.

In spite of its various potential advantages – in particular concerning economies of scale and minimum production cycle – a continuous-flow production technology is not always and necessarily the cheapest available way of production. On the one hand, plants of this sort normally involve relatively large investment outlays. Unit capital costs tend to be rather large if the equipment is less than fully utilized. On the other hand, a stop anywhere along the line can bring the whole of the line to a halt; thus, unplanned delays tend to be rather expensive. For both reasons, a continuous-flow technology can become far from economic in situations in which a steady rate of full capacity utilization is not guaranteed.

Plants embodying a discontinuous technology are very different indeed. The layout is organized into shops, whose order is by no means unique, let alone constant through time. Such factories are frequently related to the production of goods or services in small runs or in response to individual orders. Different products can be simultaneously produced, that is, the plant is not designed according to the technical transformations demanded by one specific product, but rather by groups of somewhat similar machines or tasks.

Frequent features of a discontinuous manufacturing plant are some of the following:

1. The capital equipment is less expensive and of a more general nature than that required by a continuous-flow technology.

2. There is a great deal of flexibility in the way in which a given job is performed. Given that all of the machines of a certain type can perform a particular task, the actual work load is assigned to whatever machine happens to be available. Also, similar jobs can be performed with different machines.
3. Movements of raw materials, components, subassemblies, and so on between shops become an important part of the production process, and thus it is a significant source of bottlenecks and waiting periods. The production cycle is not minimized and there is ample room for actually reducing it by carefully rearranging the physical distribution of jobs in the space.
4. Because the product is not highly standardized, on-the-job decision making is relatively important. Custom-ordered changes are normally allowed. Workers' skills in setting up the machines, preparing jigs and tools for the job, and in actually carrying out the task become very important indeed.

From these two descriptions it is obvious that continuous and discontinuous technologies would have very different consequences upon the socioeconomic structure into which they are inserted. Given the limited size of the domestic market, manufacturing firms in developing countries frequently settle for discontinuous small-batch plants of a relatively low level of automation. The choice of such a plant has a major impact upon: 1) the plant's layout; 2) the type, cost, and so on of the equipment and machinery to be installed; 3) the overall organization of production (degree and patterns of subcontracting, for example); 4) the overall number of workers; and 5) the proportion of direct to indirect labor and the like. This choice will also affect the size of the economies of scale that can eventually be captured by the firm, as well as the nature of the various technological search efforts that the manufacturing firm would find profitable to undertake through time.

In other words, not only will the physical configuration of the plant differ, but also the sources of efficiency growth (such as possibilities for capital/labor substitution and economies of scale, and possible forms of technical progress worth incorporating) will be dramatically at variance from those of a continuous-flow, highly automated manufacturing unit.

In what way can one expect domestic technological capabilities to be affected by the model of production organization? In order to answer this question one has to examine in greater detail the knowledge generation process at the individual company level.

Three categories of engineering and technical activities can be identified at the firm level that result in incremental units of new technical knowledge or information. These are: 1) product design engineering, 2) production process engineering, and 3) production planning and organization.

It should be noted that such technical activities may or may not be performed by formally organized departments or sections within the firm. The same technical activities are present even if a formal structure is absent. They are carried out by the entrepreneur himself in a small, family enterprise and are gradually decentralized and covered by specialized personnel when the firm becomes larger and more complex. The nature of such technical activities will be examined in the next few pages in order to find out exactly how it is affected by alternative forms of production organization.

Let us begin by considering the content itself of each of the technical activities carried out at the plant level.

Product Design and Specification

Being responsible for deciding what to produce, the product design department – or the design function, where there is no formal department in the firm – is responsible for the very first technical activity of any enterprise. The product design department employs different design techniques – for example, construction of prototypes and pilot plant experimentation – with the aim of attaining a final design that minimizes engineering complexity, input content, and so on for a given performance vector.

The engineering knowledge generated by this department takes the form of blueprints, formulas, and the like specifying different aspects of the product to be produced. Also, this department has the responsibility of producing incremental units of technical information on the basis of which decisions are made to upgrade, improve, or modify the currently available product design.

Economic as well as technical considerations influence the technological search efforts carried out by design personnel. On the one hand, product differentiation needs imposed by competitive pressure can frequently be identified as an inducement mechanism for product design personnel. On the other hand, new technical information – from such sources as trade journals, academic publications, and marketing surveys – could point out the need to redesign specific parts or components and/or to produce them with different raw materials or under different physical conditions. Typical of the technical search efforts carried out by product design engineering are: 1) product simplification studies, 2) standardization and normalization of parts and components, and 3) substitution of raw materials. Many of these technical efforts involve a great deal of interaction between product design engineers and members of the process engineering and industrial organization departments.

Process Engineering

The process engineering section of the firm is responsible for deciding how, by whom, and where the product should be produced. It has to select the equipment and the labor force – deciding on the size and skills of the crew – as well as the type of raw materials, components, and so on to be used in production.

It also has to work out detailed instruction sheets indicating the engineering routines to be followed, such as tolerance limits.

It is this group that will explore all potential output-stretching capabilities of the existing equipment,[12] as well as the behavior of alternative raw materials in the production process.[13] Pilot plant experimental work and time and motion and job evaluation studies constitute some of the technological search efforts normally carried out by this department. A great deal of accumulated learning underlies its activities, as it has to acquire capabilities for registering and interpreting technical information describing the behavior of the production process under different working circumstances. This involves a major step in such areas as organizational structure and the use of special equipment for the collection and processing of information.

Industrial Engineering

A third technical department with a major role in technical affairs is the one responsible for planning and control of the overall production operation. This department – normally called the industrial engineering department – issues a formal production plan stating when each action should be performed, with which machine or equipment, what should be subcontracted, what self-produced, and so on.

The industrial engineering department also decides such matters as size of batch, the machine-loading program, patterns of external subcontracting, and level of inventories, as well as plant maintenance, raw materials purchasing, and quality control.

Given the central role of the planning and control department, it has a rather wide scope for introducing changes in the everyday engineering routine of the plant.

In actual fact this department operates on the basis of a long-term plan, a short-term action program, and a control function that monitors whether or not the current operation is proceeding according to plan.

In contrast to the other two technical departments, which have very precise knowledge-generating activities whose output can be explicitly identified as a set of blueprints, production manuals, and so on, the planning and control department has a less obvious but nonetheless important knowledge-generating function. It issues on a regular basis the production plan of the firm. This is no static allocative exercise. The planning and control department fulfills a dynamic role, constantly adjusting the plant's operation to the changing signals of the marketplace.

The three technical departments described above have different roles and modi operandi depending upon whether the manufacturing plant is continuous or discontinuous.

In the case of a homogeneous commodity produced in a continuous line – automobiles and petrochemical products, for example – product design is rather inflexible and the production process tightly specified. Neither can be significantly modified. The preproduction engineering efforts, related to both product design and process engineering, are very specific, and so is the overall planning of the plant's operation. A great deal of ex ante technical work is put into balancing the production line. Time and motion studies extending right down to the level of the micromovement are required for this purpose. Given such a degree of prespecification of the production routine, the amount of ex ante technical information that has to be prepared for each position in the line is rather large. Time and motion specialists, programmers, and other such skilled personnel are employed to describe the package of technical information required for each activity.

By contrast, in discontinuous plants production planning occurs almost every time a given product is produced. There is significant scope for reducing the duration of the production cycle, which is now highly dependent on the amount of time spent on transport operations and waiting in between aisles or shops. Size of batch now becomes a crucial determinant of economies of scale. Technological search efforts to identify families of parts and components are carried out with the purpose of increasing size of batch once certain mechanical and constructional similarities among parts and components can be established. A larger batch means less time per unit in setting up the machine and, therefore, economies of scale in production.

Keeping in mind the dichotomy between continuous and discontinuous technologies, let us look at the previously described in-house knowledge-generating activities. It can be intuitively perceived that the three technical departments will have different responsibilities and will fulfill different roles in plants of one type or the other.

In the case of a continuous-flow plant, product design blueprints and production routines are available on an ex ante basis. Almost every part, component, or production subroutine is treated with equal care. Both product design and process engineering efforts are crucially different in discontinuous process plants. On the basis of what engineers call the ABC method, careful attempts are made to design some 20-30 percent of the total number of parts and components that conform to a given product design.[14]

Thus, in a discontinuous plant much more ad hoc decision making is done at the shop level, and therefore skill requirements for machine operators are significantly greater than those typically demanded by a continuous-flow plant. Skillful craftsmen with decades of experience in the actual technical secrets of each particular job substitute for job programmers and time and motion specialists.

The industrial engineering department in a discontinuous plant is responsible for issuing a machine loading program, the purpose of which is to minimize downtime between jobs and capacity underutilization resulting from imbalances between stations. By definition a continuous-flow line is balanced ex ante, the production cycle is minimized from the beginning, and no machine loading program is needed.

Other Typological Factors

Two factors in the development of indigenous technological capabilities – type of firm and production organization – have now been examined. However, these are by no means the only factors in the way domestic technological capabilities evolve through time. Other similarly important contextual factors are: size of firm, industrial field of activity, and degree of company specialization in a specific product line.

The influence of firm size upon indigenous technological development is almost but not quite self-evident. Obviously, larger firms can afford specialized engineering departments, larger R & D budgets, more expensive external advice, more complete sources of information, and the like. But what is considered large varies a great deal from one industry to another. It is quite clear that a large machine tool firm, for example – at least until very recently[15] – is tiny in comparison with plants in other metalworking sectors such as automobile or consumer durables production. Thus size needs to be considered in relation to the specific industrial field of activity.

Some industries have a greater propensity than others to adopt new, science-intensive devices, new process technologies, and new exploratory methods. The current almost explosive situation in the fields of microprocessors, optical fibers, and lasers is gradually but steadily influencing the technology and working methods of various sectors but has so far left many others almost untouched. In the area of microprocessors, the rapid fall in prices and the drastic simplification in usage – languages and machine operation in particular – is favoring a rapid diffusion of production control and product design methods almost nonexistent just five to eight years ago, and this diffusion is reaching areas as far apart as fashion design* and mechanical engineering production.[16] The next decade is almost

---

*According to a *Financial Times* report on the use of computer-aided design and computer-aided manufacturing (CAD/CAM) techniques in the clothing industry, they have recently been applied "at both the high fashion, low volume end of the market and mass produced clothes." Firms "are looking at the possibilities for developing better computer aided design and manufacturing systems." See *Financial Times*, Mar. 16, 1983, p. 33.

sure to witness an across-the-board diffusion of automation in its various forms.*

Finally, a word about the degree of company specialization. Contrary to the experience in more competitive scenarios, many of the Latin American producers examined during the course of the present enquiry opted for producing small batches of a wide mix of products rather than specializing in just one or a small number of product lines.

Obviously this has a negative influence upon downtime and efficiency in general. Not only is the possibility of economies of scale sacrificed, but also dynamic effects of the learning-by-doing type are seriously affected by this behavioral pattern. As will be shown later, it is by no means infrequent that an unduly low degree of company specialization results form policy-induced circumstances.[17]

All three of the factors thus far mentioned – firm size, field of industrial activity, and degree of specialization – undoubtedly affect both the rate and nature of the in-house technological search process that obtains at the individual firm level. Together with the two typological factors – type of firm and nature of the production process – they form the framework in which domestic technological search efforts take place.

To summarize, the development of domestic technological capabilities is associated with locally performed technological search efforts of various types. Though it is true that there is a vast array of industrial engineering areas in which almost every manufacturing plant has to become involved in order to develop firm-specific know-how – such as quality control and equipment maintenance routines – it is also clear that the type of firm, the nature of its production process, its size, its particular field of activity, and its degree of product specialization will all influence the rate and nature of the domestic technological search efforts it will find profitable to tackle through time. These are, so to speak, contour factors, even though some of them might be policy induced: for example, the choice of a discontinuous, small-batch process technology or of a rather wide

---

*Over the last few years microprocessors have rapidly diffused to countless areas such as household appliances, office equipment, and automobiles, as well as to process control devices and techniques; there is no doubt that we are witnessing the initial stages of a design revolution that is yet to come. See Chap. 2.

production mix could be a consequence of high protective tariffs or government pressure for higher domestic content.

Some of these contour factors might very well be the reason for major differences between countries in the development of domestic technological capabilities. This is bound to be the case when in one country an industry operates on the basis of continuous-flow, highly automated plants, while in another it does so on the basis of small-batch, discontinuous production processes.

MICRO AND MACROECONOMIC DETERMINANTS OF THE DOMESTIC TECHNOLOGICAL SEARCH EFFORTS

Four major sets of variables seem to affect domestic enterprises in developing countries, influencing both the rate and the direction of their technological search efforts. These are: 1) strictly micro-economic determinants resulting from the product and production technology originally available – these are firm-specific signals, mostly resulting from physical bottlenecks and imbalances in the original product design and/or plant layout; 2) forces resulting from the competitive climate prevailing in the specific market(s) to which the firm is geared; 3) macroeconomic determinants affecting firms in general, such as the rate of interest, the rate of exchange, and tariffs; and 4) new technical knowledge gained as the international technological frontier expands.

All four of these influences continuously flash out signals that reach the entrepreneurial community and trigger reactions.[18] The relative importance of each obviously changes through time in a complex way and induces firms to search in different directions or with different intensity.

Microeconomic Determinants

Manufacturing plants in developing countries very seldom constitute the result of a careful and detailed production and investment program. More often than not product designs and plant layout start off from highly unbalanced initial blueprints that are later steadily upgraded and improved. Second-hand and self-produced equipment are usually found among the initially available machinery, while the first generation of the product design frequently replicates models and specifications long outmoded on the international scene.[19] In both product design and production engineering, physical bottlenecks and imbalances appear rather frequently and cause company personnel to search for tailor-made solutions. Off-the-shelf technical knowledge might be useful as a reference point but is scarcely ready to use; what is needed is ad hoc answers that have to be worked out each time at the company level.

This point is often made in the literature. N. Rosenberg, and others have suggested that the search is problem oriented, in the sense that it is stimulated by a particular symptom that in itself defines the neighborhood in which the search effort is conducted. In Rosenberg's words, his "primary point is that most mechanical productive processes throw off signals of a sort which are both compelling and fairly obvious. Indeed, these processes, when sufficiently complex and interdependent, involve an almost compulsive formulation of problems. These problems capture a large proportion of the time and energies of those engaged in the search for improved techniques."[20]

The Competitive Climate

A second and very important set of forces affecting both the rate and the nature of the domestic technological search efforts is the competitive climate prevailing in the particular market(s) supplied by the firm. The empirical evidence collected during the course of my case studies indicates that technological search efforts are clearly influenced, both in their rate and in their nature, by the dynamics of the market's competitive atmosphere. Monopolistic situations were found to be relatively more associated with technological search efforts of the output-stretching variety. By the time the monopolistic advantage dissolves into an oligopolistic confrontation, product-differentiation search efforts and a stronger interest in cost-reducing innovations are likely to develop as well. Other things being equal, more competitive environments have been observed to lead to cost-reducing technological search efforts and product-differentiation strategies.[21]

It is important to note here that, just as individual firms change through time in aspects such as organizational structure, the ratio of skilled to unskilled personnel, and the size and complexity of the equipment and production technology they can handle, so too do markets experience significant changes in structure and competitive atmosphere. My evidence shows that in-house technological search efforts closely reflect the evolving nature of the market's competitive atmosphere.

Macroeconomic Determinants

In addition to microeconomic and market-specific variables, changes in macroeconomic parameters also influence the technological search efforts undertaken by manufacturing firms in developing countries. The magnitude of the change and the company's degree of perception seem to be crucial determinants of the pattern of reaction. Some interesting relationships have been observed during the course of the field work:

1. An increase in the cost of capital equipment has simultaneously induced entrepreneurs to postpone major investment decisions and to increase output-stretching technological search efforts aimed at extending the life cycle of specific pieces of equipment.[22] Conversely, a lower cost of capital has simultaneously enhanced the likelihood of a firm modernizing its plant with new equipment and making anticipated plant scrap decisions, thus halting output-stretching engineering efforts. Thus, government policies affecting the cost of capital seem to have had a significant impact upon both the rate and the nature of domestic technological efforts.
2. A rapid rate of demand expansion – resulting, for example, from different policy actions concerning the management of aggregate demand – frequently has triggered favorable expectations among entrepreneurs. A buoyant business climate has normally been more conducive to new construction and plant erections than to search efforts of the output-stretching variety.
3. The rate of interest also has a strong influence upon technological behavior. An increase, other things being equal, can be expected to induce search efforts directed toward the reduction of the production cycle. These could be of the product engineering sort (such as simplification of product design, standardization)

but could also involve process engineering aspects (for example, reduction of handling time between stations of a discontinuous process plant) or production planning questions (such as a more adequate management of inventories of raw materials and components).[23]

Other such examples of macroeconomic forces acting upon both the rate and the nature of the local technological search efforts undertaken by a firm are by no means difficult to find. In particular, cost and availability of R & D and engineering personnel and government support for an individual company's engineering efforts seem to be specially important in understanding the size and composition of any firm's technological search budget.

New Technical Knowledge

A fourth and final force influencing a firm's behavior in the search for new technical knowledge is the scientific and technological state of the art and worldwide changes in it over time. The rate of expansion of the best practice technological frontier and the legal, economic, and technical ease of imitation affect the technological behavior of a company both directly and indirectly.

Direct influences come from intra-industry technical progress. Various agents in any trade (engineering departments of individual firms, public R & D laboratories, process engineering and consulting firms) have as their responsibility the production of new technical knowledge. Some of it is secret and remains so for an indefinite period of time. Yet some finds its way into the public domain through professional journals, trade fairs, plant visits, and patent files.

Indirect influences derive from changes in science and technology in general. Advances in, say, fundamental physics, metallurgy, or biology eventually find their way into better machines or new product designs. The current irruption of computer-aided design and computer-aided manufacturing (CAD/CAM) systems across the industrial spectrum constitutes one of the more impressive indirect influences upon company behavior, this time deriving from developments in microelectronics, solid state physics, computer sciences, and so forth.

The extent of a company's exposure to new information and the ability and skills of the technical personnel in pursuing the most promising leads will determine the relative weight and influence of these variables upon company technological behavior. Channels for the reception of technical information must be set up and operated, and the inflowing information must be decoded in a way that reflects its importance to the company. Both of these are highly firm specific and will to a large extent decide the degree of the firm's technological currentness. A firm's exposure to the flow of information and its specific capability to screen and decode it in relation to its particular needs appear as yet a scarcely understood subject. Trade journals, technical letters exchanged with equipment suppliers, clients, licenses, trade fairs, and visits to other plants appear to be the relevant units of measurement in this field. As far as I am aware, no one has yet systematically tried to undertake a study of this problem.

It is important to note at this point that the relative role and incidence of each of the four sets of variables influencing the rate and direction of the domestic technological search efforts will certainly change over time. Periods of overall macroeconomic stability, steady expansion of demand, and low uncertainty will presumably make firms pay more attention to company-specific variables and search for a more balanced utilization of the available plant production capacity, for a more thorough debugging of product designs, for example. On the other hand, periods of higher macroeconomic instability probably reduce the relative pay-off of domestic technological search activities of any sort while simultaneously increasing the expected benefits of alternative, short-term courses of action.

THE DEVELOPMENT OF DOMESTIC TECHNOLOGICAL CAPABILITIES AS A SEQUENTIAL PROCESS

The structure of the learning process that obtains at the company level as a consequence of the four variables just discussed must now be examined.

Industry and individual firm studies seem to indicate that an evolutionary sequence normally prevails as far as in-house technological search efforts are concerned. Firms appear gradually to develop technological capabilities of their own, proceeding stepwise, from the more immediate, less risky to the more long-term, more complex areas. However, the meaning of these terms depends upon both circumstances and technical skills already possessed by the firm's engineering staff. Circumstances here refer both to given typological conditions, such as company field of activity, and also to policy-induced reality. In other words, it is to be expected that the evolutionary sequence to be followed by a family metalworking firm, created around the mechanical ingenuity of a self-made technician and producing an array of different metallurgical products, would significantly differ from that of a continuous-flow petrochemical plant producing a homogeneous commodity under the supervision of a staff of university-trained chemical engineers. Similarly, that evolutionary sequence will also probably be sensitive to public policy decisions such as those related to tariffs and subsidization schemes of various sorts that lower the cost of capital for the individual entrepreneur.

For the sake of argument let us examine two extreme cases of sequential development of indigenous technological capabilities, beginning with a mechanical engineering firm producing a mix of metallurgical products in a discontinuous, small-batch plant. This is a type frequently found throughout the Latin American metalworking sector.

Technological search efforts in the area of product design seem to appear rather early in the technical history of many of the firms of this sort examined during the course of my work. Only a few years after start-up these firms seem to begin developing in-house technical skills related to product design engineering. On the basis of these skills they first adapt and improve the original versions of the produced item(s) and then start playing product differentiation strategies as part of their competitive behavior.[24] The life cycle of industrial products and the relatively low incentive to search for cost reduction innovations, given the rather extreme degree of protection prevailing in most of these cases, appear as major explanations for the early development of product design engineering capabilities. Many of these firms begin with indigenous technological efforts in the area of product design well before they can exhibit significant technical strength in other technological areas. Prototypes and plant experimentation for product design purposes seem to appear well before time and motion studies or other such tools of production planning and engineering are employed by in-house technical personnel.

This should not be taken to mean that technological search in areas related to process engineering is entirely absent during those initial years of company operation. Rudimentary forms of search are almost invariably present during the start-up period. Also, substitution of one raw material for another, the introduction of new or improved products, and so on are activities that necessarily call for some limited amount of search concerning both the production process and the organization of production.

A certain discontinuity was uncovered in the technical history of most of the firms examined somewhere during the second decade of operation. This frequently involved a major change of attitude concerning process engineering and production planning activities and was associated with a new approach toward questions of quality control, limits of tolerance, preventive maintenance, and other such technical matters. In many cases the change of attitude was related both to a significant reorganization of the firm's administrative structure (with the creation of a number of new departments such as quality control, research and development, and tooling) and to a major increase in the size and complexity of the firm's output mix. Both changes call for a rather different way of handling inventories, machine loading programs, quality control, and the manipulation of parts and pieces within the plant, for example. Process engineering and organizational skills generated in an informal way during the initial years of company operation are normally found to be insufficient at this point, thus indicating the need for a radical change in organizational structure, in data gathering efforts, and so forth.

A new approach toward engineering efforts frequently obtains after such discontinuity and is noticeable in a drastic increase in the ratio of indirect to direct labor inputs associated with the creation of a number of new technical functions. The overall production operation is now run according to a central plan and is carefully monitored with an eye to out-of-equilibrium situations.

Consider now a second possible case: a continuous-flow process plant producing a homogeneous product, say a petrochemical raw material, under the control of a team of university-trained chemical engineers. Furthermore, assume it to be the domestic subsidiary of an MNC. Given that the commodity is standardized, there is little flexibility in the product design area. Since the plant is a subsidiary of an MNC, a great deal of the product formulation effort is probably made by its parent company. Yet there is some room for maneuver in terms of product mix. Plants of this sort normally operate on the basis of a certain vector of nationally available raw materials – whose prices, qualities, and so on are normally known ex ante – and a certain demand vector that needs to be satisfied. It is here that some freedom of action does obtain in terms of product choices and of product formulation, but, on the whole, continuous-flow process plants are less demanding in terms of product engineering activities and relatively more so in areas of process engineering.[25] The application of plant and process optimization methods is rather frequent in firms of this kind, and this calls for the early creation of a strong process engineering department capable of tinkering with pilot plant experiments that might yield detailed information about the behavior of the production process under varying operating circumstances – such as different raw materials or sources of energy. Such information constitutes the basis for the design of locally adapted process engineering routines capable of substituting for the original ones brought from abroad.

Plants of this sort do not normally develop strong production planning departments either. The machine-paced nature of the transformations and unitary operations they carry out do not normally leave room for much flexibility. Instead, many of these firms develop their own project engineering departments capable of designing new production plants, supervising their construction and start-up and, finally, handing them down to production engineers once they have been partially debugged and brought close to normal operation. It is interesting that some of these project engineering departments have been the origin of various now independent process engineering companies whose main job is that of selling turnkey plants to third parties, both nationally and in other Latin American countries. These departments frequently start by licensing basic engineering technology from major international subcontractors and supplying their own detailed engineering know-how. Gradually they substitute foreign for local technology even in areas of basic engineering, and this presupposes the indigenous development of various engineering and scientific fields of activity.

DOMESTIC TECHNOLOGICAL CAPABILITIES AND PROTECTION

From the empirical evidence collected during the research program and examined above, the development of indigenous technological capabilities in developing countries may be described as a sequential process highly sensitive to the type of firm, the nature of the production process, the firm's field of activity and ownership, and so on, as well as the competitive atmosphere and the macroeconomic parameters within which it operates.

On the basis of these results it is possible to give concrete meaning – and, therefore, a useful one for the purposes of public policy – to the concept of indigenous technological learning. In spite of its being central to the debate on infant industry protection, this concept has remained unexplored in the received literature.[26]

Let us think of a successful indigenous learning sequence as one that enables a firm (or an industry) to create and develop a complete set of in-house engineering skills – that is, product design skills, process engineering capabilities, and production planning and organization skills – so as to be able to compete on its own both in domestic and in international markets.

My research shows that a complete learning sequence in this sense calls for the incorporation into the firm of a large number of engineers and technicians of different specialties and training levels, ranging from design engineers to time and motion specialists and including draftsmen and process engineers. The research also shows that two or more decades have often been necessary for a given firm to complete the cycle.

Now, neither all the countries included in my enquiry nor all types of firms and forms of production organization encountered during the field work appear to have been equally successful in completing the learning sequence. Firstly, there are those firms (or industries) in which the production process is of the continuous type, organized in line, both in developed countries (DCs) and in the Latin American cases examined by my team. Within this group, firms exhibit an extremely uneven performance across countries, Brazilian companies being much more successful than those operating in, say, Argentina or Colombia. Secondly, manufacturing plants both in Latin America and in DCs are organized as a discontinuous array of shops producing small batches of quasi-standard goods or tailor-made individual orders. This form of production organization is structurally different from the first, and therefore the analysis of the case for protecting it in developing countries in order to attain its efficient consolidation should be carried out in terms of different variables and arguments than in the former case.

Finally, there is a third case that needs examination as well: those industries that are characterized in developing countries by small-batch, discontinuous plants but in developed countries by continuous-flow, in-line establishments.

Let us first look at the case of continuous-flow production. Of the several cases of this sort studied in the program, two Brazilian firms stand out as possible success stories. They are Romi and Metal Leve, producing conventional lathes and pistons respectively. Both have 1) moved toward international plant scales, 2) dramatically expanded their R & D efforts, approaching the international technological frontier, and 3) increased their exports, actively competing in markets in developed countries.

There appear to be three reasons why these firms have attained better long-term performance than comparable companies in, say, Argentina or Colombia. Firstly, the larger size and dynamism of the Brazilian domestic market permitted them to count on at least two decades of protected expansion upon which to base their own consolidation. Secondly, and perhaps as a consequence, both firms made a special point of erecting plants on an international scale. Such a choice – absent in most other countries in the region – permitted them to capture large economies of scale. Finally, it is striking that both operate in industrial fields in which the world's technological frontier did not experience significant changes until very recently, thus permitting a gradual narrowing of the gap that separated them from international standards.

In other words, in those cases in which the size of the domestic market has permitted the erection of plants of international scale – that is, those cases that do not exhibit static diseconomies of scale vis-à-vis establishments in the developed world – and in which the world's technological frontier has not experienced significant jumps for a long period of time, an infant industry protection policy, systematically maintained for at least two decades, seems to have induced the development of competitive national enterprises based upon sound indigenous technological capabilities.

It should be noted that my analysis[27] presupposes as given the state of the art at the international level. If we accept that the world's state of the art is in fact changing – in the above cases, numerically controlled lathes partially substituting for conventional lathes and new alloys substituting for conventional raw materials in the production of pistons – the situation becomes somewhat more complex and calls for a dynamic analysis that would explicitly incorporate observed changes in the state of the art. There are two possibilities here.

Firstly, if the new product or production process emerging at the world technological frontier is not a perfect substitute for the old one – that is, if enough demand for the mature product (or process) is likely to exist – its future production need not encounter difficulties even if the state of the art changes quite significantly. In other words, the local firm could continue to exploit its achieved technological learning even if it is now in the context of a mature product. In terms of our Brazilian examples, while the demand (national or international) for lathes and pistons of the conventional type persists, the case for having protected the learning period seems to be a valid one.

Secondly, if sooner or later the mature product (or process) disappears and is completely replaced by the new one, the domestic technological capabilities arising from the production of the mature product (or use of the mature process) may not be the most appropriate ones for successfully keeping pace with (or even leading) the movement of the international technological frontier. To do so may well require a different set of skills that the mature product (or process) may not have provided (or even required) or may have provided only partially.

Something of this sort seems to be taking place in at least one of the previously mentioned Brazilian firms in its transition from conventional lathes – an area in which the firm has given evidence of having successfully completed its learning process – to numerically controlled lathes, where the company has now felt the need of a licensing agreement with a large multinational firm, which will supply technical knowledge in electronic fields that the previous, entirely mechanical production process did not provide. It is clear that the case is somewhat more complicated than the one argued by classical economists. Should society take upon itself to protect a second, or even a third, round of indigenous learning in order to prevent the technological gap from widening once again, or should the returns from the first learning sequence finance the dynamics of the evolving process? This is a major theoretical and policy issue that has thus far received little attention in the literature.

None of the other large-scale in-line metalworking plants in the sample can be regarded as an equally successful case. In each, numerous forms of indigenous technological learning were identified. In spite of that, however, they are all far from being able to compete successfully on their own with large-scale international metalworking companies.

The situation changes significantly when discontinuous plants that produce in small batches or custom-made orders are examined.

Let us first take the case of tailor-made, individual-order items, such as complex boilers, heat exchangers, and turbines, whose production is carried out in plants of a discontinuous nature in both developed and developing countries. This type of metalworking plant is less subject to conventional economies of scale and enjoys a certain degree of natural protection arising from the tailor-made nature of the product or of the customer. It is in this alternative framework that the prospects for protectionist policies aimed at inducing the "naturalization" of this kind of industry in Latin America must be examined.

While market size appeared to be crucial in large-scale, continuous-flow manufacturing plants, in this case in-house engineering design capabilities and the availability of modern and updated metalworking machinery become decisive. So is the availability of human skills adequate to operate and maintain sophisticated design and production equipment.

The likelihood of successfully developing metalworking plants of this sort in Latin American countries depends upon the availability of a relatively large supply of skilled technicians and engineers, as well as specialists in software, computing sciences and the like. Plants of this kind have already been set up in Argentina, Brazil, Mexico, and it is in those countries that the likelihood of their future successful development seems brighter. Two points are worth making here. On the one hand, there is clear evidence that significant structural changes are already under way in such countries as Brazil and Argentina in the development of an indigenous software. One can expect a learning sequence to emerge there that resembles that of mechanical or process engineering during the 1960s and 1970s. On the other hand, the world's technological frontier in this field is moving ahead at a rate that is going to make the NICs' competitive situation much more difficult to maintain in the decades ahead.

A further observation appears to be justified at this point. A major customer for most of the products of this type of metal-working firm is the public sector itself through its various public agencies responsible for energy, transportation, and so on. Hence the public sector could provide stable long-term demand that would ensure close-to-capacity utilization of the available skills and machinery. This is an issue of public policy that should be carefully considered by Latin American governments interested in supporting the development of domestic technological capabilities in firms of this sort while simultaneously being responsible for a large fraction – over 50-60 percent in many of the countries in the region – of current investment programs.

The advisability for Latin America of a third type of industry warrants examination. This is the type in which production is carried out in small-batch, discontinuous plants in Latin America but in much more highly automated, continuous-flow plants in the developed world. As in the two previous cases, governments should consider carefully the extent to which protection should be offered to such firms in order to bring about their naturalization to the local environment.

In addition to conventional relative factor price differentials that can, per se, partially or totally offset the underlying disadvantages, two other factors need to be taken into account in this case.[28] On the one hand, reference has already been made to the existence of a certain degree of "natural protection" in many manufacturing fields. This originates in localization advantages and/or in technological adaptation to the domestic environment. The competitive edge of the local firm does not, however, finish there. In recent years considerable progress has been made in Latin America in the field of production planning and industrial organization (like the reorganization of plant layout in cells or technology groups), which make it possible to capture significant economies of scale even within the framework of a discontinuous production organization.[29]

The simultaneous impact of a certain degree of natural protection and economies of scale through the successful utilization of modern, computer-based production planning and organization methods may very well result in a successful naturalization of manufacturing plants producing in small lots, even in industrial fields where firms in developed countries operate with large-scale plants.

Of course, this is not a matter of logical necessity. Since we are comparing two different production organizations – continuous-flow production, as in developed countries, vis-à-vis discontinuous, shop-wise production, as in developing countries – the actual efficiency gap between them appears to be of an empirical nature with significant differences across industries. The feasibility of successfully naturalizing these industries in Latin America cannot be ruled out a priori, but in those cases in which the scale differentials are really large, an attempt to force the naturalization through subsidization will necessarily lead to socially suboptimal situations.

CONCLUSION

To sum up, manufacturing production can be carried out by at least three forms of industrial organization, that is, continuous-flow, small-batch, and tailor-made production. The feasibility of successfully nationalizing a specific type of production by inducing its installation and development through protection depends upon: 1) the initial gap between the production organization chosen by plants in developed and developing countries; 2) the rate of indigenous technological progress attained by plants in developed and developing countries; 3) conventional factor price differentials; and 4) the degree of natural protection enjoyed by the product.

From the Latin American empirical evidence, it seems that, except for specific examples inherent in the Brazilian case, large-scale in-line production of standardized products lags behind international standards. The relative success of domestic manufacturing production appears to be much greater in industrial fields that produce individual orders or small series, particularly where the underlying disadvantages are counteracted by large domestic engineering efforts in product design and production planning and organization, and by sophisticated computer-based production technology.

NOTES

1. K. Arrow, "Economic Welfare and the Allocation of Resources for Invention," in *The Rate and Direction of Inventive Activity* ed. R. Nelson (Princeton, N.J.: Princeton University Press, 1962), p. 609; J. Stiglitz, "On the Microeconomics of Technical Progress," Monografía de Trabajo no. 32, Programa BID/CEPAL/PNUD de Investigaciones en Temas de Ciencia y Tecnología en América Latina (Buenos Aires, April 1979); R. Nelson and S. Winter, *An Evolutionary Theory of Economic Change* (Cambridge: Harvard University Press, Belknap Press, 1982).

2. M. Peck, with the collaboration of S. Tamura, "Technology," in *Asia's New Giant* ed. H. Patrick and H. Rosovsky (Washington, D.C.: Brookings Institution, 1976).

3. J. Katz and E. Ablin, "From Infant Industry to Technology Exports: The Argentine Experience in the International Sale of Industrial Plants and Engineering Works," Monografía de Trabajo no. 14, Programa BID/CEPAL/PNUD de Investigaciones en Temas de Ciencia y Tecnología en América Latina (Buenos Aires, October 1978); P. O'Brien, *Technology Exports from Developing Countries: The Case of Argentina* (New York: UNIDO, March 1981); R. Soifer, "Estudio de las exportaciones de tecnología argentinas" (Buenos Aires: INTAL, July 1982), mimeograph; C. Dahlman and F. Sercovich, "Exports of Technology from Semi-Industrial Economies," paper presented at the meeting of the American Economic Association, New York, December 1982; J. Katz and B. Kosacoff, "Direct Foreign Investment of Argentine Industrial Enterprises" (Buenos Aires: Institute for Research and Information on Multinationals, 1982); C. Diaz Alejandro, "Foreign Direct Investment by Latin Americans," in *Multinationals from Small Countries* ed. T. Agmon and C. P. Kindelberger (Cambridge, Mass.: MIT Press, 1977); L. Wells, "La internacionalización de firmas de los países en desarrollo," *Revista de Integración Latinoamericana*, no. 14 (Buenos Aires: INTAL, June 1977); S. Lall, *Third World Multinationals* ed. S. Lall (London: Macmillan, 1983).

4. Book I in A. Smith's, *Wealth of Nations*, deals extensively with these topics in the famous case of the pin-making trade.

5. R. Nelson, "Research on Productivity Growth and Productivity Differences: Dead Ends and New Departures," *Journal of Economic Literature*, 19 (September 1981), pp. 1031, 1037.

6. H. Nogueira da Cruz, "Evolucão tecnológica no setor de máquinas de processar cereais. Um estudo do caso," Monografía de Trabajo no. 39, Programa BID/CEPAL/CIID/PNUD de Investigaciones sobre Desarrollo Científico y Tecnológico en América Latina (Buenos Aires, July 1981); A. Castaño et al., "Etapas históricas y conductas tecnológicas en una planta argentina de máquinas herramienta," Monografía de Trabajo no. 38, Programa BID/CEPAL/CIID/PNUD de Investigaciones sobre Desarrollo Científico y Tecnológico en América Latina (Buenos Aires, January 1981); M. Turkieh et al., "El cambio tecnológico en la industria venezuelana de maquinaria agrícola. Estudios de case," Monografía de Trabajo no. 52, Programa BID/CEPAL/CIID/PNUD de Investigaciones sobre Desarrollo Científico y Tecnológico en América Latina (Buenos Aires, July 1982).

7. H. Nogueira da Cruz, "Evolucao tecnológica."

8. J. Berlinski et al., "Basic Issues Emerging from Recent Research on Technological Behavior of Selected Latin American Metalworking Plants," Monografía de Trabajo no. 56, Programa BID/CEPAL/CIID/PNUD de Investigaciones sobre Desarrollo Científico y Tecnológico en América Latina (Buenos Aires, August 1982), chap. 2; J. Katz, "Cambio tecnológico en la industria metal-mecánica latinoamericana. Resultados de un Programa de Estudios de Casos," Monografía de Trabajo no. 51, Programa BID/CEPAL/CIID/PNUD de Investigaciones sobre Desarrollo Científico y Tecnológico en América Latina (Buenos Aires, July 1982).

9. Concerning the relative success of domestically owned pharmaceutical firms in Argentina, see J. Katz, *Oligopolio, empresas nacionales y firmas multinacionales. El caso de la industria farmaceutica*. Siglo XXI (Buenos Aires: 1974); D. Chudnovsky, "The Challenge by Domestic Enterprises to the Transnational Corporation's Domination: A Case Study of the Argentine Pharmaceutical Industry," *World Development* vol. 7, no. 1 (London: January 1979), pp. 45-58.

10. In relation to this topic the case studies by D. Sandoval concerning Forjas de Colombia, a public sector metalworking company, and C. Dahlman's enquiry into the technological history of Brazil's USIMINAS, a steel producing firm, provide interesting contrasting evidence. See D. Sandoval et al., "Analisis del desarrollo industrial de Forjas de Colombia. 1961-1981," Monografía de Trabajo no. 50, Programa BID/CEPAL/CIID/PNUD de Investigaciones sobre Desarrollo Científico y Tecnológico en América Latina (Buenos Aires, June 1982); C. Dahlman et al., "From Technological Dependence to Technological Development: The Case of the Usiminas Steel Plant in Brazil," Monografía de Trabajo no. 21, Programa BID/CEPAL/PNUD de Investigaciones en Temas de Ciencia y Tecnología en América Latina (Buenos Aires, October 1978).

11. Dahlman et al., "From Technological Dependence to Technological Development"; Dahlman and Sercovich, "Exports of Technology."

12. The notion of the "output-stretching" flexibility embodied in one specific plant design has been explored by P. Maxwell and M. Teubal in the content of the steel industry. See their "Capacity-Stretching Technical Change: Some Empirical and Theoretical Aspects," Monografía de Trabajo no. 36, Programa BID/CEPAL/PNUD de Investigaciones sobre Temas de Ciencia y Tecnología en América Latina (Buenos Aires, December 1980); J. Katz, "Productivity, Technology, and Domestic Efforts in Research and Development," Monografía de Trabajo no. 13, Programa BID/CEPAL/PNUD de Investigaciones sobre Temas de Ciencia y Tecnología en América Latina (Buenos Aires, July 1978).

13. J. Katz, "Productivity, Technology, and Domestic Efforts in Research and Development"; C. Dahlman et al., "From Technological Dependence to Technological Development"; L. A. Pérez, and J. Pérez, "Análisis microeconómico de las características del cambio tecnológico y del proceso de innovaciones. El caso de Furfural y Devivados S.A., Mexico," Monografía de Trabajo no. 20, Programa BID/CEPAL/PNUD de Investigaciones sobre Temas de Ciencia y Tecnología en América Latina (Buenos Aires, June 1978).

14. "Introducción al estudio del trabajo," pp. 86 ff.

15. J. Hartley, "Japanese Show Huge Gains with Unmanned Operation," in *Iron Age Metalworking International* (PA: Chilton, 1982).

16. See the whole issue of *Iron Age Metalworking International*, vol. 21, no. 7 (1982).

17. D. Sandoval in his study of Forjas de Colombia, J. Berlinski in his analysis of Perkins Argentina, as well as other authors in recent writings give evidence in this respect. See Sandoval et al., "Forjas de Colombia," and Berlinski, "Innovaciones en el processo."

18. The degree of entrepreneurial perception of such signals will certainly affect the pattern of reaction. See H. Schwartz, "Perception, Judgement and Motivation in Decision Making: Hypotheses Suggested by a Study of Metalworking Enterprises in Argentina, Mexico and the United States," (March 1980), mimeograph; Katz, "Cambio tecnológico."

19. Castaño et al., "Etapas históricas"; Nogueira da Cruz, "Evolucao tecnológica."

20. N. Rosenberg, *Perspectives on Technology* (Cambridge: Cambridge University Press, 1976).

21. J. Katz, "Productivity, Technology and Domestic Efforts in Research and Development"; J. Katz, "Domestic Technology Generation in LDCs: A Review of Research Findings," Monografía de Trabajo no. 35, Programa BID/CEPAL/PNUD de Investigaciones sobre Desarrollo Científico y Tecnológico en América Latina (Buenos Aires, November 1980). Concerning the influence of market structure upon technological behavior, see F. M. Scherer, *Industrial Market Structure and Economic Performance* (Chicago: Rand McNally & Co., 1970), chap. 15; R. Nelson and S. Winter, "In Search of a Useful Theory of Innovation," *Research Policy* 6 (1977): p. 36.

22. P. Maxwell, "First-Best Technological Strategy in an 'Nth-Best' Economic Context. A Case Study of the Evolution of the Acindar Steel Plant in Rosario, Argentina," Monografía de Trabajo no. 16, Programa BID/CEPAL/PNUD de Investigaciones sobre Temas de Ciencia y Tecnología en América Latina (Buenos Aires, April 1978); Maxwell and Teubal, "Capacity-Stretching Technical Change."

23. Castaño et al., "Etapas históricas." For a more general statement concerning this particular effect, see Katz, "Cambio tecnológico."

24. For the case of metalworking companies producing agricultural equipment, see A. J. Berlinski, "Innovaciones en productos y aprendizaje (El caso de una planta argentina de implementos agrícolas)," Monografía de Trabajo no. 60, Programa BID/CEPAL/CIID/PNUD de Investigaciones sobre Desarrollo Científico y Tecnológico en América Latina (Buenos Aires, September 1982); Turkieh et al., "El cambio tecnológico"; Castaño et al., "Etapas históricas."

25. Consider, for example, the results obtained by F. Sercovich in his study of petrochemical plants, or those by L. A. Pérez in his study of a Mexican plant producing furfural. See F. Sercovich, "Design Engineering and Endogenous Technical Change. A Microeconomic Approach Based on the Experience of the Argentine Chemical and Petrochemical Industries," Monografía de Trabajo no. 19, Programa BID/CEPAL/PNUD de Investigaciones en Temas de Ciencia y Tecnología en América Latina (Buenos Aires, October 1978); Pérez and Pérez,

"Analysis Microeconómico de las características del Cambio Tecnológico del proceso de inovaciones."

26. It is rather unfortunate that contemporary writings on infant industry protection are not based upon a careful micro exploration of learning sequences in different manufacturing setups. Quite the contrary, a lot of unproven "conventional wisdom" seems to guide them. See B. Balassa and M. Sharpston, "Export Subsidies by Developing Countries: Issues of Policy," World Bank Reprint Series no. 51 (Geneva, November 1977); B. Balassa, "Export Incentives and Export Performance in Developing Countries: A Comparative Analysis," World Bank Reprint Series no. 59 (Tübingen, 1978); J. Bhagwati and T. N. Srinivasan, "Trade Policy and Development," World Bank Reprint Series No. 90 (Washington, D.C.: 1978). An entirely different perspective can be obtained from the work of Nelson and Winter, "In Search of a Useful Theory"; Rosenberg, *Perspectives on Technology*.

27. In relation to the classical case for infant industry protection, see J. S. Mill, *Essays on Some Unsettled Questions of Political Economy* (1874; reprint; Fairfield, N.J.: Augustus M. Kelley, 1974); M. C. Kemps, "The Mill-Bastable Infant Industry Dogma," *Journal of Political Economy* (February, 1960).

28. A clear analytical framework examining the extent to which conventional factor price differentials can partially (or totally) offset the underlying scale disadvantages has been presented by H. Pack in "The Capital Goods Sector in LDCs. A Survey," (Washington, D.C.: April 1980), mimeograph.

29. Arn, *Group Technology*. Concerning its use in the Latin American region, see A. Mercado and P. Toledo, "El cambio tecnológico en una empresa mexicana productora de máquinas para el vidrio y el plastico," Monografía de Trabajo no. 57, Programa BID/CEPAL/CIID/PNUD de Investigaciones sobre Desarrollo Científico y Tecnológico en América Latina (Buenos Aires, September 1982); Castaño et al., "Etapas históricas."

REFERENCES

A. Aráoz. "Consulting and Engineering Design Organizations in Developing Countries." December 1977. Mimeograph.

E. A. Arn. *Group Technology*. New York: Springer-Verlag, 1975.

K. Arrow. "Economic Welfare and the Allocation of Resources for Invention." In *The Rate and Direction of Inventive Activity*, edited by R. Nelson. Princeton, N.J.: National Bureau of Economic Research, Princeton University Press, 1962.

B. Balassa. "Export Incentives and Export Performance in Developing Countries: A Comparative Analysis." World Bank Reprint Series no. 59. Tübingen, 1978.

B. Balassa and M. Sharpston. "Export Subsidies by Developing Countries: Issues of Policy." World Bank Reprint Series no. 51. Geneva, November 1977.

J. Baranson and R. Roark. "Trends in North-South Transfer of High Technology." Chapter 2, this volume.

J. Berlinski. "Innovaciones en el proceso y apprendizaje (El caso de una planta argentina de motores)." Monografía de Trabajo no. 59, Programa BID/CEPAL/CIID/PNUD de Investigaciones sobre Desarrollo Científico y Tecnológico en América Latina. Buenos Aires, September 1982.

A. J. Berlinski. "Innovaciones en productos y apprendizaje (El case de una planta argentina de implementos agrícolas)." Monografía de Trabajo no. 60, Programa BID/CEPAL/CIID/PNUD de Investigaciones sobre Desarrollo Científico y Tecnológico en América Latina. Buenos Aires, September 1982.

J. Berlinski, H. Nogueira da Cruz, D. Sandoval, and M. Turkieh. "Basic Issues Emerging from Recent Research on Technological Behavior of Selected Latin American Metalworking Plants." Monografía de Trabajo no. 56, Programa BID/CEPAL/CIID/PNUD de Investigaciones sobre Desarrollo Científico y Tecnológico en América Latina. Buenos Aires, August 1982.

J. Bhagwati and T. N. Srinivasan. "Trade Policy and Development." World Bank Reprint Series no. 90. Washington, D.C., 1980.

G. E. Box. "Some General Considerations in Process Optimization." *Journal of Basic Engineering*, March 1970.

A. Castaño, J. Katz, and F. Navajas. "Etapas históricas y conductas tecnológicas en una planta argentina de máquinas herramienta." Monografía de Trabajo no. 38, Programa BID/CEPAL/CIID/PNUD de Investigaciones sobre Desarrollo Científico y Tecnológico en América Latina. Buenos Aires, July 1982.

D. Chudnovsky. "The Challenge by Domestic Enterprises to the Transnational Corporation's Domination: A Case Study of the Argentine Pharmaceutical Industry." *World Development* vol. 7, no. 1. London: Pergamon Press, January 1979, pp. 45-58.

C. Dahlman and F. Sercovich. "Exports of Technology from Semi-Industrial Economies." Paper presented at the meeting of the American Economic Association, New York, December 1982.

C. Dahlman. "From Technological Dependence to Technological Development: The Case of the Usiminas Steel Plant in Brazil." Monografía de Trabajo no. 21, Programa BID/CEPAL/PNUD de Investigaciones en Temas de Ciencia y Tecnología en América Latina. Buenos Aires, October 1978.

C. Diaz Alejandro. "Foreign Direct Investment by Latin Americans." In *Multinationals from Small Countries*, edited by T. Agmon and C. P. Kindelberger. Cambridge, Mass.: MIT Press, 1977.

Financial Times, March 16, 1983, p. 33.

C. Freeman. "Chemical Process Plant: Innovation and the World Market." *National Institute Economic Review*, August 1968.

J. Hartley. "Japanese Show Huge Gains with Unmanned Operation." In *Iron Age Metalworking International* vol. 21, no. 7. Pa.: Chilton, 1982.

S. Jacobson. "Trends and Implications of Automation in the Engineering Industry – A Preliminary Analysis." Research Policy. Lund: Lund Institute, University of Lund, 1982. Mimeograph.

J. Katz. "Cambio tecnológico en la industria metal-mecánica latinoamericana. Resultados de una Programa de Estudios de Casos." Monografía de Trabajo no. 51, Programa BID/CEPAL/CIID/PNUD de Investigaciones sobre Desarrollo Científico y Tecnológico en América Latina. Buenos Aires, July 1982.

____. "Domestic Technology Generation in LDCs: A Review of Research Findings." Monografía de Trabajo no. 35, Programa BID/CEPAL/CIID/PNUD de Investigaciones sobre Desarrollo Científico y Tecnológico en América Latina. Buenos Aires, November 1980.

____. *Oligopolio, empresas nacionales y firmas multinacionales. El caso de la indutria farmaceutica*. Siglo XXI, Buenos Aires, 1974.

____. "Productivity, Technology and Domestic Efforts in Research and Development." Monografía de Trabajo no. 13, Programa BID/CEPAL/PNUD de Investigaciones sobre Temas de Ciencia y Tecnología en América Latina. Buenos Aires, July 1979.

J. Katz and B. Kosacoff. "Direct Foreign Investment of Argentine Industrial Enterprises." Buenos Aires: Institute for Research and Information on Multinationals, 1982. Mimeograph.

J. Katz and E. Ablin. "From Infant Industry to Technology Exports: The Argentine Experience in the International Sale of Industrial Plants and Engineering Works." Monografía de Trabajo no. 14, Programa BID/CEPAL/PNUD de Investigaciones en Temas de Ciencia y Tecnología en América Latina. Buenos Aires, October 1978.

M. C. Kemps. "The Mill-Bastable Infant Industry Dogma." *Journal of Political Economy*, February 1960.

S. Lall. *Third World Multinationals*. London: Macmillan, 1983.

P. Maxwell. "First-Best Technological Strategy in an 'Nth-Best' Economic Context. A Case Study of the Evolution of the Acindar Steel Plant in Rosario, Argentina." Monografía de Trabajo no. 16, Programa BID/CEPAL/PNUD de Investigaciones sobre Temas de Ciencia y Tecnología en América Latina. Buenos Aires, April 1978.

P. Maxwell and M. Teubal. "Capacity-Stretching Technical Change: Some Empirical and Theoretical Aspects." Monografía de Trabajo no. 36, Programa BID/CEPAL/PNUD de Investigaciones sobre Temas de Ciencia y Tecnología en América Latina. Buenos Aires, December 1980.

A. Mercado and P. Toledo. "El cambio tecnológico en una empresa mexicana productora de máquinas para el vidrio y el Plástico." Monografía de Trabajo no. 57, Programa BID/CEPAL/CIID/PNUD de Investigaciones sobre Desarrollo Científico y Tecnológico en América Latina. Buenos Aires, September 1982.

J. S. Mill. *Essays on Some Unsettled Questions of Political Economy*. 1874 reprint. Fairfield, N.J.: Augustus M. Kelley, 1974. Essay 1.

I. Miyata. "Economical Methods for Building a Flexible Manufacturing System." In *Metalworking Engineering and Marketing* vol. 5, no. 2. March 1983.

Organizacion International del Trabajo. *Introduccion al estudio del trabajo*. Ginebra: OIT, 1966.

R. Nelson. *The Rate and Direction of Inventive Activity*. Princeton, N.J.: National Bureau of Economic Research, Princeton University Press, 1962.

____. "Research on Productivity Growth and Productivity Differences: Dead Ends and New Departures." *Journal of Economic Literature*, September 1981.

R. Nelson and S. Winter. *An Evolutionary Theory of Economic Change*. Cambridge: Harvard University Press, Belknap Press, 1982.

H. Nogueira da Cruz. "Evolucão tecnológica no sector de máquinas de processar cereais. Um estudo de case." Monografía de Trabajo no. 39, Programa BID/CEPAL/CIID/PNUD de Investigaciones sobre Desarrollo Científico y Tecnológico en América Latina. Buenos Aires, July 1981.

——. "In Search of a Useful Theory of Innovation." *Research Policy*, 1977.

P. O'Brien. *Technology Exports from Developing Countries: The Case of Argentina*. New York: UNIDO, March 1981.

H. Pack. "The Capital Goods Sector in LDCs. A Survey." Washington, D.C., April 1980. Mimeograph.

M. Peck, with the collaboration of S. Tamura. "Technology." In *Asia's New Giant*, edited by H. Patrick and H. Rosovsky. Washington, D.C.: Brookings Institution, 1976.

L. A. Pérez and J. Pérez. "Análisis microeconómico de las características del cambio tecnológico y del procoso de innovaciones. El caso de Furfural y Derivados S.A., México." Monografía de Trabajo no. 20, Programa BID/CEPAL/PNUD de Investigaciones sobre Temas de Ciencia y Tecnología en América Latina. Buenos Aires, June 1978.

N. Rosenberg. *Perspectives on Technology*. Cambridge: Cambridge University Press, 1976.

D. Sandoval, Mick, L. Outerman, and L. Varamillo. "Análisis del desarrollo industrial de Forjas de Colombia. 1961-1981." Monografía de Trabajo no. 50, Programa BID/CEPAL/CIID/PNUD de Investigaciones sobre Desarrollo Científico y Tecnológico en América Latina. Buenos Aires, June 1982.

F. M. Scherer. *Industrial Market Structure and Economic Performance*. Chicago: Rand McNally & Company, 1970.

H. Schwartz. "Perception, Judgement and Motivation in Decision Making: Hypotheses Suggested by a Study of Metalworking Enterprises in Argentina, Mexico and the United States." March 1980. Mimeograph.

F. Sercovich. "Design Engineering and Endogenous Technical Change. A Microeconomic Approach Based on the Experience of the Argentine Chemical and Petrochemical Industries." Monografía de Trabajo no. 19. Programa BID/CEPAL/PNUD de Investigaciones en Temas de Ciencia y Tecnología en América Latina. Buenos Aires, October 1978.

R. Soifer. "Estudio de las exportaciones de tecnología argentinas." Buenos Aires: INTAL, July 1982. Mimeograph.

J. Stiglitz. "On the Microeconomics of Technical Progress." Monografía de Trabajo no. 32, Programa BID/CEPAL/PNUD de Investigaciones en Temas de Ciencia y Tecnología en América Latina. Buenos Aires, April 1979.

M. Turkieh, A. Pirela, J. M. Martínez, and P. Levin. "El cambio tecnológico en la industria venezuelana de maquinaria agrícola. Estudios de case." Monografía de Trabajo no. 52, Programa BID/CEPAL/CIID/PNUD de Investigaciones sobre Desarrollo Científico y Tecnológico en América Latina. Buenos Aires, July 1982.

L. Wells. "La internacionalizacion de firmas de los países en desarrollo." *Revista de Integración Latinoamericana* no. 14. Buenos Aires, INTAL, June 1977.

J. Zlotsky. *Seguimiento y control de estudios, proyectos y obtas de ingeniería*. Buenos Aires: Editorial Ballestra, 1976.

6

Reflections on The Republic of Korea's Acquisition of Technological Capability

Larry E. Westphal,
Linsu Kim, and
Carl J. Dahlman

INTRODUCTION

South Korean industrialization is unusual not only in its remarkable pace but in other significant respects as well. It thus is a fruitful area for studying industrial development, as is demonstrated by what has been written on the trade aspect of the Republic of Korea's industrial strategy. But there has been little study of – and even less written about – the technological development accompanying its industrialization. Our purpose in this chapter is to document the technological aspect of South Korean industrial strategy.

In our previous attempts to convey what we have learned about South Korea's technological development to colleagues, we have been frustrated by the fact that economists typically do not think

---

The research underlying this chapter forms part of an ongoing project, "The Acquisition of Technological Capability," sponsored by the World Bank (Ref. No. 672-48). Westphal had the primary responsibility for writing the chapter. Kim worked mainly on the last three sections; financial support for research assistance from the Social Science Research Council enabled him to undertake the supplementary case history research that is reported in the section "Exports in Relation to Technological Development." Dahlman worked mainly on the first two sections. Bruce Ross-Larson edited the manuscript for publication.

This chapter had its origin in a related paper presented at the Conference on International Technology Transfer: Concepts, Measures, and Comparisons; sponsored by the Social Science Research Council, Subcommittee on Science

of industrialization as technological development. That may be the result of the absence of a satisfactory analytical framework for thinking in these terms. We thus find it necessary to establish a conceptual foundation for discussing the South Korean experience. The first two sections provide this foundation, which – though largely derived from our reflections on South Korea – we believe is more generally valid. South Korea's technological development is described in the next section. Because of the importance of the export-led character of its industrialization, we explore then the interactions between technological development and export activity. Findings and speculations are summarized in the final section.

TECHNOLOGICAL DEVELOPMENT IN RELATION TO INDUSTRIALIZATION

Why is it that economists typically do not associate technological development with the industrialization of developing countries? Perhaps because invention, the central aspect of global technological development, plays only a minor part in the process. Most technology introduced into developing countries is transferred in one way or another from industrially more advanced countries. But because industrialization adds to the variety of products produced and processes used in a country, it surely does involve technological development in the sense of gaining mastery over products and processes that are new to the local economy. The minor role of invention simply means that much technological development consists of assimilating foreign technology.

Another possible explanation for the neglect of technological development in industrialization is that assimilation often seems to

---

and Technology Indicators. We are grateful to the participants at the conference for useful comments and discussion and want to acknowledge a special debt to Richard Nelson. We also wish to acknowledge Alice Amsden's contributions as co-author (with Westphal and Kim) of the original paper. Scheduling conflicts precluded her collaborating on this chapter.

The authors are solely responsible for the views and interpretations expressed in this paper. In particular, these views and interpretations should not be attributed to the World Bank, to its affiliated institutions, or to any individual acting on behalf of these organizations.

be characterized as being automatic and without cost. If this were correct, assimilation would not merit much attention. But it is not accomplished by passively receiving technology from overseas. It requires investments in understanding the principles and use of technology, investments reflected in increased human and institutional capital. These have a big effect on productivity. And they have a big effect on possibilities for adapting technology to suit local circumstances better.

A third possible explanation for the scant attention paid to technological development is that assimilation is seen as being not only absolute but also unitary and discrete. But rather than being absolute, assimilation has an aspect of proficiency. And in addition, the choices associated with investments in assimilation have an aspect of complexity that follows from the character of technology as a compound system of interrelated elements. If this complexity is forgotten, as it often appears to be, the nature of assimilation is trivialized.

The complexity of choices associated with investments in assimilation – complexity inherent in the separability of technology into its many elements – has extremely important implications. Most important is that a country does not have to possess all the capabilities needed to provide each of the various elements of technology. All the elements of technology can in principle be obtained – individually or in a variety of combinations – from foreign sources. It is in this respect that South Korea provides a particularly interesting and instructive case. Its experience during the 1960s and 1970s shows how a country can successfully assimilate technology in piecemeal fashion.

International trade in the elements of technology means, too, that there is no necessary relationship between industrial development and technological development – the former seen in the size and composition of industrial output and the latter in the local capability to provide various elements of technology. Certainly there is no necessary relationship over short periods. That is why many paths of local technological development can be taken to reach the same level of industrialization. Those paths can differ in the mode of technology transfers as well as in the extent and nature of investments to assimilate technology – to acquire the capabilities to provide the various elements. There naturally are differences in costs and benefits for different modes of transfer. But these may be of secondary importance, since they can be overcome by tailoring investments in assimilation to offset them.

How, then, is technological development to be assessed in relation to industrial development? Any survey of a country's past technological development can at best yield imprecise results because of the limited usefulness of readily available information. Many indirect modes of technology transfer, such as ideas conveyed through the open technical literature, are not easily traced. Many forms of investment in assimilating technology, such as those associated with the accumulation of experience, also go unrecorded. Except in rare circumstances, costs cannot be assessed precisely, even through detailed analyses of case histories of firms and sectors. Benefits are at least as difficult to assess because of the difficulties created by the likelihood of externalities. Contemporary attempts to uncover the costs and benefits associated with different paths of technological development are further hampered by the paucity of adequate case histories.

In sum, current surveys of technological development in industrializing countries can only address descriptive questions: What elements of technology have been assimilated and in what sequence? Where did they come from? How were they first obtained and then absorbed? Which of them were adapted, in what respects, and how? These are the questions we address in this chapter.

THE NATURE OF TECHNOLOGICAL CAPABILITY[1]

Analysis of technological development requires a conceptual framework that distinguishes the possible choices associated with investments in assimilation. In building such a framework, we have found it most useful to focus not on technology but on technological capabilities distinguished by the functions served. In this section we will provide a functional classification of technological capabilities and discuss how these relate to trade in the elements of technology.

The term *technology* refers to a collection of physical processes that transform inputs into outputs, to the specifications of the inputs and the outputs, and to the social arrangements that structure the activities involved in carrying out these transformations. Technology thus is the practical application of technological knowledge, which is what underlies and is given usable expression in technology. Such knowledge has technical and transactional elements, the former relating to product characteristics and physical processes, and the latter to social arrangements. These social arrangements include various kinds of market and contractual relationships among entities as well as organizational modes and procedural methods to regulate an entity's internal operations.

Technological capability is the ability to make effective use of technological knowledge. It is the primary attribute of human and institutional capital.* It inheres not in the knowledge that is possessed but in the use of that knowledge and in the proficiency of its use in production, investment, and innovation.† Because of the complexity of technology, there are many distinct technological capabilities, classifiable in numerous ways, each corresponding to a different way of distinguishing the aspects of technological knowledge and its application. We sketch here a classification by broad areas of application.

Functional Classification of Capabilities

Technological capabilities are separable into three broad areas: production, investment, and innovation.‡ The first capability is for

---

**Institutional capital* refers to the know-how used to combine human skills and physical capital into systems for delivering want-satisfying products. In terms of Hall and Johnson's conceptual framework, institutional capital includes system-specific and firm-specific technology plus some elements of general technology. G. R. Hall and R. E. Johnson, "Transfers of United States Aerospace Technology to Japan," in *The Technology Factor in International Trade*, ed. Raymond Vernon (New York: National Bureau of Economic Research, distributed by Columbia University Press, 1970), pp. 305-58.

†The identification of technological capability as something distinct and worthy of attention conforms to Salter's separation of technological knowledge and technology. He uses the terms *technical knowledge* and *techniques of production*. W. E. G. Salter, *Productivity and Technical Change* (Cambridge: Cambridge University Press, 1960), pp. 13 ff.

‡Hayami and Ruttan use an analogous typology in discussing transfers of agricultural technology. They distinguish transfers of material, design, and capability, which as they define them imply the attainment of production, investment, and innovation capability. Yujiro Hayami and Vernon W. Ruttan, *Agricultural Development: An International Perspective* (Baltimore: Johns Hopkins Press, 1971), pp. 175 ff.

operating productive facilities, the second is for expanding capacity and establishing new productive facilities, and the third is for developing technologies. Proficiency in production capability is reflected in technical or x-) efficiency and in the ability to adapt operations to changing market circumstances. Proficiency in investment capability is reflected in project costs and in the ability to tailor project designs to suit the special circumstances of the investment. Proficiency in innovation capability is reflected in the ability to develop technologies that are less costly and more effective.

The boundaries separating these broad areas of technological capability recognize differences that originate in the specialization of activities applying technological knowledge. Table 6.1 provides a list of the principal activities associated with production and investment capability. Much technological knowledge is tacit – it resides as much in minds as in manuals. That is why experience is critical for becoming proficient in each of these activities. But the scope for transferring capability gained through experience in one activity to other activities is limited.[2] For example, the understanding required for repairing and maintaining physical capital has very little in common with that required for optimizing the scheduling of production. For another example, basic engineering requires highly specialized knowledge of the relevant core processes, while detailed engineering activities, such as designing architectural and civil works, require other specialized knowledge.

Moreover, the experience gained in operating, say, a steel mill has little to do with (in the sense that it is not sufficient for) establishing a new steel mill. The differences between the activities involved and the specialized knowledge used are simply too great. In fact, many elements of the experience gained in the operation of one steel mill may not transfer to the operation of another using a somewhat different technology. But the transferability varies across industries. For example, experience operating textile factories may suffice for the establishment of similar factories since textile machinery is available in standardized designs.

Innovation capability pertains to the activities of conceiving and implementing changes in the technical and transactional elements of technology. Included are all the activities spanning invention to innovation, ranging from radical (major) new departures to

TABLE 6.1
Elements of Production and Investment Capability

Production Capability

Production management – to oversee the operation of established facilities

Production engineering – to provide the information required to optimize the operation of established facilities, including:

- Raw material control – to sort and grade inputs, seek improved inputs
- Production scheduling – to coordinate production processes across products and facilities
- Quality control – to monitor conformance with product standards and to upgrade them
- Trouble-shooting – to overcome problems encountered in the course of operation
- Adaptations of processes and products – to respond to changing circumstances and to increase productivity

Repair and maintenance of physical capital – according to regular schedule or when needed

Marketing – to find and develop uses for possible outputs and to channel outputs to markets

Investment Capability

Manpower training – to impart skills and abilities of all kinds

Preinvestment feasibility studies – to identify possible projects and to ascertain prospects for viability under alternative design concepts

Project execution – to establish or expand facilities, including:

- Project management – to organize and oversee the activities involved in project execution
- Project engineering – to provide the information needed to make technology operational in a particular setting, including:
  - Detailed studies – to make tentative choices among design alternatives
  - Basic engineering – to supply the core technology in terms of process flows, material and energy balances, specifications of principal equipment, plant layout
  - Detailed engineering – to supply the peripheral technology in terms of complete specifications for all physical capital, architectural and engineering plans, construction and equipment installation specifications
- Procurement – to choose, coordinate, and supervise hardware suppliers and construction contractors

continued

Table 6.1, Continued

Embodiment in physical capital – to accomplish site preparation, construction, plant erection, manufacture of machinery and equipment
Start-up of operations – to attain predetermined norms

Notes: The elements listed under production capability refer to the operation of manufacturing plants, but similar activities pertain to the operation of other types of productive facilities as well. In turn, our use of "production engineering" departs from conventional usage in that we use the term far more broadly to include all of the engineering activities related to the operation of existing facilities. In our usage, the term encompasses "product design" and "manufacturing engineering" as these terms are generally used in reference to industrial production. See the entries under these headings in the *McGraw-Hill Encyclopedia of Science and Technology* (New York: McGraw-Hill Book Company, 1977).

Source: Larry E. Westphal and others, "Exports of Capital Goods and Related Services from the Republic of Korea," World Bank Staff Working Paper, no. 629 (Washington, D.C.: The World Bank, 1984).

incremental (minor) improvements in existing technology.* Basic research has no direct relationship with innovation capability because its objective is to gain knowledge for its own sake. But there is an obvious and important indirect relationship because basic research feeds into applied research and development, either as the source of new knowledge or as the locus of human and institutional capital accumulation. Innovation capability is directly related to applied research and development, the objectives of which are to obtain knowledge with specific commercial applications and to translate it into concrete operational form. It needs to be recognized, however, that some development activities, such as the creation of pilot plants, also involve production and investment capabilities.

As the examples of textiles and pilot plants suggest, the boundaries that separate the three broad areas of capability – production, investment, and innovation – are fuzzy, particularly in adaptation. In industrializing countries as elsewhere, the successful

*It can be argued that the introduction of existing technology into a new environment requires capabilities that are not unlike some of those involved in innovation, as it is commonly defined. But in our framework this form of "local innovation" entails investment capability rather than innovation capability.

use of technology in production is known to require continuous adaptation involving the activities associated with production capability (see Table 6.1).* Adaptation is important because the optimum utilization of technology is rarely achieved at the outset. Not only is the scope for optimizing within given circumstances almost limitless, but circumstances are constantly changing in input and output markets.[3]

Adaptations are not costless – they require the allocation of resources to purposive technological efforts. The changes are typically implemented through trial-and-error testing of modifications that is like applied research, even though it often does not involve inputs from specialized agents of R & D, either within the firm or outside it. Similarly, adaptation often involves elements of investment capability. This is true because adaptations are frequently embodied in minor changes or additions to physical capital. In some cases these adaptations take place through a process that encompasses all aspects of investment capability, even though the amount invested is small.

More generally, experience-based technological efforts to adapt production technology can provide part of the understanding necessary to carry out some – but not all – of the activities involved in investment and innovation. For example, plant engineers may acquire some capability in plant design and spare parts production from experience in breaking bottlenecks, maintaining equipment, and solving production problems. But it is unlikely that they will acquire a capability in basic plant design, capital goods manufacture, or the creation of radically new technologies. The background and experience needed to carry out many of these activities is different, and the relevant capabilities tend to be developed in specialized entities such as process engineering firms, capital goods producers, and technological research institutes. In turn, the accumulation of local experience in carrying out investments can provide some of the understanding that is relevant to innovation – for example,

---

*See Dahlman and Westphal, "Technological Effort in Industrial Development." They conclude from the available evidence that "the cumulative sequence of (minor, incremental) technological changes following the initiation of a (successful) new activity (in an industrializing country) may have a greater impact on the productivity of employed resources than that produced by its initial establishment" (pp. 113-17).

knowledge of basic engineering or the ability to embody technology in capital goods. But innovation often involves highly specialized technical knowledge that goes beyond what is needed in relation to investment.

To summarize, the boundaries between areas, though fuzzy, show that specific investments are required to assimilate distinct capabilities. These investments often involve experience in adapting existing technology. Moreover, the differentiating characteristics of the various technological capabilities are loosely reflected in the institutions that organize and coordinate economic activity.

But patterns of specialization and exchange shift both within and among entities, many of which are newly evolved, as existing capabilities are strengthened and additional capabilities are added. The changes are generally, but not always, in the direction of increasing specialization. Where warranted by – among other determinants – the extent of the market, some activities become the domain of specialized units within individual entities or of specialized entities. The latter include a wide variety of firms that provide various marketing, consulting, and engineering services; firms that engage in the manufacture and construction of physical capital; and institutions that perform educational and training services. Producing firms necessarily possess some production capability, aspects of which may be concentrated in differentiated marketing, engineering, and other departments. Nonetheless, such firms may often rely on outside entities for services related to marketing, quality control, process optimization, plant and equipment maintenance, and the like. Moreover, producing firms almost always rely on outside entities for at least some aspects of investment capability: feasibility studies, architectural and specialized engineering designs, and fabrication of physical capital, for example. In turn, specialized innovation capability is found in R & D departments and in separate R & D establishments.

Shifts in patterns of specialization and exchange are important because they are the way that the transactional elements of technology change, and rarely can the full promise of technical changes be realized without changes in these elements.[4] Such institutional change is accomplished through investments that are embodied in the form of organizational structures, codified knowledge and procedures, and customs that govern behavior within and among entities. Thus there is good reason to distinguish institutional from human and physical capital and to say that technological capability resides in both human and institutional capital.

Relationship of Technological Capabilities to Trade in the Elements of Technology

The specialization and exchange in modern economies are such that few entities are wholly integrated systems with all the technological capabilities relevant to their establishment and operation. Instead, technological capabilities are deployed across entities through market transactions in the elements of technology. These transactions take many forms involving goods, services, and information. And they take place between countries as well as within them. Historically, transactions between countries have increased tremendously in volume and diversity as part of the phenomenon of growing global economic interdependence.

For analytical purposes it helps to distinguish several broadly defined elements present – either singly or in combination – in international trade in the elements of technology.

- *Technological knowledge*: the information about physical processes and social arrangements that underlies and is given operational expression in technology.
- *Technical services*: the activities of translating technological knowledge into the detailed information required to establish or operate a productive facility in a specific set of circumstances.
- *Embodiment activity*: the activities of forming physical capital in accord with given and complete design specifications.*
- *Training services*: the activities of imparting the skills and abilities that are employed in economic activity.
- *Management services*: the activities of organizing and managing the operation of productive facilities, the implementation of investment projects, and the development of product and process innovations.
- *Marketing services*: the activities of matching the capacity of productive facilities to existing and latent market demands.

---

*Reasons for separately distinguishing embodiment activity are given below.

The first of these elements is technological knowledge in whatever form it is available without processing or translation to tailor it to (a different set of) specific circumstances. Knowledge in this form is often obtained without payment – that is, without a market transaction explicitly meant to secure it. Most of the other elements are sometimes also obtained without payment, but much less frequently. What distinguishes these is that they consist of activities that reflect the application of technological knowledge in establishing and using productive facilities. Most transactions involving the elements of technology combine technological knowledge with one or more such activities. These activities are the same ones as were discussed in relation to the various technological capabilities; the elements listed above simply cluster them in a different way. The importance of distinguishing the activities from the technological knowledge that underlies them rests in the need for a clear framework to analyze comparative advantage in relation to trade in the elements of technology.

Trade in the elements of technology takes many transactional modes that can serve numerous objectives. Licensing, subcontracting, technical agreements, management contracts, marketing arrangements, turnkey project contracts, direct foreign investment, and trade in capital goods are only some of the forms. Part of this trade simply provides complementary services without any real flow of technology.* Marketing services provided under international subcontracting are a case in point, though they are often combined with technical services that yield an inflow of technology. Several forms of this trade involve packages of more than one element in addition to technological knowledge. Direct foreign investment and turnkey project contracts are obvious examples. In turn, while all forms can play an important role in developing local capabilities, this role is not assured for any of them. Even training activities, which have the ostensible objective of imparting technological capabilities, do not always achieve their goals. Moreover, to the extent that imported elements are not readily available from local sources, imports are more obviously meant – at least in their initial effect – to substitute for or complement local capabilities.

*In this respect such trade might preferably be called trade "involving" – rather than "in" – the elements of technology, but we prefer the single label that better describes most of the trade we are concerned with.

Trade in the elements of technology makes it possible to develop most industries through any one of many possible combinations of local and foreign capabilities. At one extreme is the autarkic creation (or re-creation) of technology by locally providing all of the requisite elements of technology through accumulating the relevant technological capabilities. This extreme is likely to prove both very costly and very time-consuming even if use is made of freely available foreign technological knowledge. It nevertheless guarantees the acquisition of at least rudimentary proficiency in the relevant technological capabilities. At the other extreme, an industry can be established and operated on the basis of foreign capabilities with no local accumulation of capability. This sometimes happens, for example, with direct foreign investment in an enclave where indigenous involvement is limited to an unskilled labor force. This extreme may prove an effective way of generating employment and earning foreign exchange in the short to medium term, but it does not contribute to technological development. Between these extremes are many possibilities, some of which are illustrated in the following sections on South Korea's experience.

Although trade in the elements of technology can play an important role in developing local capabilities, the relationship is not a simple one. Trade in these elements transfers the elements, not the capabilities to provide them, certainly not as a direct (or immediate) result of their being imported. To illustrate: Turnkey projects are a frequent vehicle for establishing new industries in developing countries. These projects package all the elements required to establish a production facility and train people in its operation. Rarely do they transfer any of the investment capabilities. Instead, they are intended to transfer production capability.

The relationship is further complicated by the fact that capabilities that are ostensibly intended to be acquired are sometimes not acquired. This is true because technological development can occur only as the result of purposive effort to assimilate technology. There is abundant and compelling evidence on this point from numerous instances of unsuccessful attempts. In turn, capabilities that are typically not expected to be acquired, given the nature of the transaction, can be acquired as the result of purposive effort. Examples are given in the following sections on South Korea.

It is impossible to discuss technological development meaningfully without reference to the objectives sought. Trade in the elements of technology is an important consideration in assessing these objectives. Much technological development is import substitution to replace foreign capability with indigenous capability in activities related to local production and investment. The benefits of this technological development can – and often do – extend beyond simple import substitution to include the ability to adapt technology. Moreover, such development can increase exports, including exports of the elements of technology, as it has in all semi-industrial economies. And there are possible externalities for technological development in other areas to be considered as well. But the accumulation of human and institutional capital in specific technological capabilities also has costs, which include the investment of time and resources needed to accumulate specific capabilities. In addition, as initial experience is being gained, there may be higher costs and greater risks in using local capabilities.

Technological development is not to be seen as having the objective of progressive import substitution for *all* of the elements of technology. Costs and benefits cannot be evaluated without taking account of other opportunities for technological development that are necessarily forgone (or postponed) by dedicating scarce high-level manpower to one area. Particularly in the face of continual global technological development, efficiency in local technological development implies continued imports of many elements of technology, though the pattern of imports shifts as indigenous capabilities are developed.

To conclude: trade in the elements of technology is a critical dimension of any strategy for technological development, which in turn is an important correlate of industrial strategy. Industrial strategies are often discussed simply in terms of the sectoral composition of industry and the market orientation of industrial activity vis-à-vis foreign trade. One of our purposes in this chapter is to demonstrate that technological strategy can fruitfully be discussed in the same terms as industrial strategy, in terms of the sequence of choosing the capabilities to be acquired. As should be apparent from the foregoing discussion, sequencing is not simply a matter of the particular industries in which technological capabilities are acquired. It involves broadening – in part by adding capabilities to accomplish new technological activities in established industries. It involves deepening by achieving greater proficiency and increased differentiation of existing capabilities. And it involves increasing discrimination in the selection of elements of technology that can be efficiently supplied through local development. In turn, technological capabilities can be employed not only in simple import substitution but also in adaptation and in exporting and can be the base from which additional capabilities are acquired. The following sections discuss South Korea's industrial and related technological development from this perspective.

THE REPUBLIC OF KOREA'S INDUSTRIAL AND TECHNOLOGICAL DEVELOPMENT

South Korea's rapid sustained industrialization dates from the mid-1960s. Most of our discussion is correspondingly focused on the second half of the 1960s and the 1970s. But the heritage from before 1965 needs to be kept in mind when one contemplates how the South Koreans were able to make such effective use of modern technology after 1965. Thus we begin with a brief historical sketch, then turn to the past two decades and examine South Korea's technological development, and conclude the section with a discussion of the acquisition of technological capability.

Historical Perspective

Modern industrialization in Korea began in the colonial period, when the Japanese government managed the peninsula's economy as an integral part of its empire. Manufacturing growth during the colonial period was rapid and extensive, but it depended heavily on the Japanese. Nonetheless, Koreans apparently acquired, mostly on the job, substantial knowledge about how to operate modern industries. There is not universal agreement, however, on just how much human capital was built up during the colonial period. Suh emphasizes the "imposed" enclave nature of industrialization during the colonial period and concludes that the colonial bequest of human capital was negligible.[5] But Mason and others emphasize the "'demonstration effect' of exposure to modern technology and forms of organization" and conclude that the colonial bequest was considerable.[6] We are not sure just how much South Korea's subsequent industrialization owes to the colonial period. All we know for certain is that its colonial experience was unlike that of most other developing countries in that it involved considerable industrialization.

The Korean economy suffered tremendous disruption at the end of World War II when its ties to Japan were severed and the peninsula was divided into two political entities.[7] In what now is the Republic of Korea, or South Korea, manufacturing production in 1945 was substantially less than one-fifth of its level in 1940. But in the light of circumstances at the time, what is really remarkable is that the South Koreans were able, with relatively little foreign managerial or technical assistance, to operate nearly half the manufacturing plants that existed in 1944: there was no sector in which they were unable to produce at least something. By 1948, with assistance from the United States – access to raw materials, replacement parts, and technical help – the South Koreans were operating facilities to produce a wide variety of manufactured goods, including shoes, textiles, rubber tires, basic steel shapes, and such engineering products as pumps, bicycles, tin cans, and ball bearings.

Because of the Korean War, adjustments to dislocation continued to dominate economic activity until the mid-1950s. A respectable but not outstanding rate of industrial expansion was achieved during the latter half of the 1950s, with import substitution for light manufactured and nondurable consumer goods playing the major role. A more important development during this period was the tremendous expansion of education, to which U.S. aid contributed.[8] By 1960 the Republic of Korea had achieved universal primary education, nearly universal adult literacy, and rapidly growing enrollment rates at all levels above the primary level. U.S. aid also financed overseas education and training for thousands of South Koreans. In addition, Americans helped the South Korean military learn modern concepts and techniques of management and organization, as well as how to operate and maintain all types of machinery and equipment. For almost all the male labor force, military service, which was universal, seems to have been an important source of skill formation and general experience in an organization having many characteristics of modern industry.

The South Koreans also gained some industrial competence from their relationship with the United States. Important channels for the direct transfer of industrial technology included the inflow of technical advisers and a modest volume of project assistance. The U.S. military was another channel. Its local procurement program afforded producers in a number of sectors with occasions for assisted learning-by-doing to meet exacting product specifications. Among those benefiting from military purchases were construction contractors, plywood producers, and tire makers – all would later become major exporters.

Large-scale U.S. economic assistance continued through the 1960s. Beginning in the late 1950s, however, there was a gradual shift from grant aid to concessional lending. In addition, the South Korean government began to promote inflows of foreign resources of all kinds from increasingly diversified sources. In the early 1960s, as opportunities for easy import substitution rapidly diminished, industrial growth began to falter. A number of attempts were made at policy reform and economic liberalization in the first half of the 1960s, as policy makers came to accept that rapid economic development depended on export-oriented industrialization and a greater effort to mobilize domestic and foreign resources. These attempts culminated in 1964 and 1965, when a number of reforms were successfully implemented.[9]

Industrial Development after 1965[10]

After 1965 the growth of manufactured exports and the rise in domestic demand fueled a rate of industrialization that was much faster than before. The average annual rate of growth of manufacturing production during 1965-81 was twice the 11 percent rate during 1955-65. Underlying the acceleration of growth in manufacturing, the share of exports in gross manufacturing output increased to roughly one-third in 1981 from less than one-tenth in 1965 – it was nil in 1955. During 1965-81 the ratio of exports to gross national product (GNP) more than quadrupled, and the share of GNP originating in the manufacturing sector more than doubled. Manufacturing exports became increasingly diversified, so that by 1981 South Korea was a major exporter of steel, footwear, machinery and transport equipment, and various manufactures of metal and nonmetallic minerals – in addition to textiles, clothing, and plywood, which had led the initial growth of exports.

There was rapid structural change in manufacturing after 1965. Though the share of textiles, apparel, and leather in manufacturing value added (at 1975 prices) was the same (18 percent) in 1966 and in 1981, the share of food, beverages, and tobacco declined from 37 to 20 percent. Roughly offsetting this decline were increases in the shares of chemical products, including petroleum refining (from 15 to 23 percent), basic metals (from 5 to 10 percent), and machinery and transport equipment (from 13 to 20 percent). South Korea's chemical industry had its start in the early 1960s when large-scale petroleum refining and fertilizer production were introduced. The first petrochemicals complex came on line around 1970. The most notable single event in the development of the basic metals industry was the construction of South Korea's integrated steel mill in the early 1970s. The initial growth of the machinery sector was centered in electrical appliances and electronics. Production of electrical machinery grew far more rapidly than that of nonelectrical machinery throughout the period. In transport equipment the most remarkable achievement has been in shipbuilding, following the establishment of one of the world's largest shipyards in the early 1970s.

Until the mid-1970s the government's strategy largely operated on the accumulation of production capability. Indeed, before the shift in strategy, government policy discriminated against domestic investment capability, most notably by favoring imported capital goods. This policy bias was reversed in the early 1970s, at about the time that the *Heavy and Chemical Industry Plan*, a long-term plan covering the decade to 1981, was published. Among other things, this plan called for the rapid buildup of capacity to manufacture capital goods, particularly the fabricated structural elements and heavy equipment used in industrial plants producing basic intermediate goods and in power generation and transmission and other social overhead facilities.

As will be discussed later, exports of capital goods and related services – in other words, exports of elements of technology – burgeoned after the mid-1970s. But the manner in which they grew was not foreseen by the plan. Indeed, the plan was heavily focused on import substitution, reflecting the government's intention not to increase the economy's dependence on export activity. But very soon thereafter the government abandoned its intention. This, together with gradual recognition of the extent of its overambitiousness, led to several revisions of the plan. In the light of the country's emerging exports of elements of technology, successive revisions became increasingly focused on export activity.

It was also during the mid-1970s that the government began to give serious priority to technological development,* and export activity likewise became an integral part of its efforts to promote the acquisition of technological capability more generally. Separate legislative acts to promote the technological development of producing firms and to encourage the development of engineering services firms included the provision of incentives for exports of the elements of technology. These measures were supplemented by others designed to foster the education and training of qualified personnel in various technical fields and to establish an infrastructure of scientific and technological institutes designed to serve industry.

We turn now to a brief examination of South Korea's reliance on formal technology transfers, starting with the role of direct foreign investment (DFI). Such investment has played only a minor part in the country's industrialization. There was no DFI between 1945 and 1960, when the first legislation controlling nongrant inflows of foreign capital was promulgated. The first instance of DFI in the period after World War II was in 1962. By the end of 1981, the government had approved 693 instances (or cases) of DFI in manufacturing. The cumulative gross inflow amounted to roughly U.S. $1.25 billion. This figure may be compared with cumulative total gross investment in manufacturing over the same period, U.S. $22.7 billion, to show the relative magnitude of DFI.†

A big part of DFI was approved on condition that it involve exports. Nonetheless, it made very little contribution to the expansion of exports during 1962-71 – that is, during the period when export expansion replaced import substitution as the primary engine of industrial development. Indeed, South Korea's first free-trade zone explicitly designed to attract foreign participation in exports

---

*We know that the Ministry of Science and Technology and the Korean Institute of Science and Technology were created in the mid-1960s and were active before the mid-1970s. But we judge that the real shift in priorities came well after their establishment.

†Notwithstanding the relative unimportance of DFI, other foreign capital inflows have been very important as a source of the investment finance needed to achieve rapid industrialization. For example, though the share of capital inflows in total investment has fallen steadily over time, it was 20 percent even as late as 1972-76.

was not established until 1970.* A disproportionate inflow of DFI occurred in 1972-76 (see Table 6.2), much of it oriented toward exporting. DFI intended for offshore production became relatively less important after 1976.

The composition of DFI during 1962-81 is compared with that of value added in Table 6.2. For comparability, values for both are cumulative over the 20 years. The principal sectors for DFI included textiles and apparel, chemicals (synthetic fibers and resins plus petroleum refining and other chemicals), and electrical and nonelectrical machinery. Most DFI in textiles and apparel and a large share of it in electronics were related to production for export and played almost no role in technological development. In the former sector, much of the inflow consisted of the relocation of small plants from Japan to take advantage of low wages. In the latter sector, much was for offshore assembly with very little spin-off to local producers. In contrast, DFI has been a particularly important vehicle for technological development in the establishment of much of the chemicals sector and, more recently, of major elements of the electrical and nonelectrical machinery sectors. DFI has also contributed to technological development in the basic metals sector, but there is no foreign equity in the integrated steel mill.

Formal flows of disembodied technology via other modes also appear to have been rather modest. The cumulative value of technical assistance in manufacturing from bilateral and multilateral sources during 1962-81 was well under U.S. $100 million, as was the cumulative value of technical consultancy by private parties. The latter was heavily concentrated in the chemical and machinery (both electrical and nonelectrical) sectors. Table 6.2 also shows the composition of licensed technology imports. The total number of approved manufacturing technology imports during 1962-81 was 1,840; royalty payments over the same period totaled U.S. $565 million. The volume of licensed imports was rather modest until the mid-1970s. Thereafter the increased reliance on licensing can be explained by the accelerated development of the technologically more advanced industries in recent years. Licensing has been an important source of technology transfer in much the same industries as DFI – chemicals, basic metals, and machinery.

---

*Investments in it and a second zone established later accounted for roughly 10 percent of the cumulative inflow of DFI by the late 1970s.

International comparisons give a particularly fruitful perspective on the Republic of Korea's pattern of reliance on different modes of technology transfer. Table 6.3 compares South Korea with four other semi-industrial economies – Argentina, Brazil, India, and Mexico – using the best information that we have been able to obtain. Additional cross-country information cited by Westphal and others supports the indication given by the first two blocks of comparative data that DFI and disembodied technology inflows via commercial channels have by no means been relatively large in South Korea.[11] In turn, the third block of data confirms what knowledge of the economy would lead one to suspect, namely that the country's reliance on imported capital goods has, in contrast, been relatively large. Imports of capital goods were more than 20 percent of the value of investment in South Korea throughout the 1970s. The closest country of the other four in this respect was Mexico, with ratios of 11-14 percent during the 1970s.

South Korea's dependence on imported capital goods should be seen as a result of specialization within the capital goods sector and of the demands of a rapidly growing and diversifying industrial sector, rather than as the result of failure to develop a capital goods sector. While it is true that the growth of nonelectrical machinery production failed to keep pace with the growth of industrial output, capital goods production nonetheless grew during 1962-81. The capital goods sector dates back to the colonial period. Over time, all the important metalworking processes, such as casting and machining, were assimilated by South Korean firms and used in copying many types of imported equipment, with the designs subsequently modified on the basis of experience to make them more appropriate to local circumstances. But the capability to design and produce capital goods was oriented toward the more labor-intensive segments of those industries that had a relatively long history in the country. Most export industries used imported equipment extensively, as did most new industries established under government incentives.

Insight about how the South Koreans were able effectively to assimilate technology comes from the comparative data on human capital formation in Table 6.3. What stands out about the educational pattern are the high proportion of postsecondary students abroad, the high secondary enrollment rate, and the high percentage of engineering students among postsecondary students. Even more remarkable is the rapid growth in numbers of scientists and engineers,

TABLE 6.2
Proprietary Transfers of Technology by Sector and Plan Period

	Foreign investment		*Licensed technology*		*Production*
	Approvals granted	*Amount invested*	*Approvals granted*	*Royalty payments*	*Value added*
Percentage distribution of cumulative values by sector, 1962-81					
Food, beverages, and tobacco	2.6	3.6	2.5	1.1	24.0
Textiles, apparel, and leather	9.7	7.2	1.6	1.1	19.0
Pulp and paper products	0.4	0.1	0.8	1.5	2.5
Pharmaceuticals	2.0	1.6	2.9	0.4	*
Synthetic fibers and resins	14.4	30.9	2.7	4.5	2.8
Petroleum refining and other chemicals	1.1	8.2	18.7	36.8	20.2
Cement and ceramic products	3.2	1.6	3.0	2.4	5.1
Basic metals	9.1	6.7	9.7	11.4	7.1
Nonelectrical machinery	17.2	8.9	31.6	21.3	2.3
Electrical machinery	24.8	22.4	19.5	12.3	9.3
Transport equipment	1.0	5.0	3.0	3.3	5.3
Other manufacturing	14.4	3.9	4.0	3.7	2.3
Total manufacturing	100.0	100.0	100.0	100.0	100.0

Percentage distribution by plan period					
1962-66	2.0	2.5	1.6	0.1	4.4
1967-71	20.1	6.7	14.2	2.7	11.5
1972-76	56.4	47.3	22.5	18.4	28.3
1977-81	21.5	43.5	61.8	78.8	55.8
Total 1962-81	100.0	100.0	100.0	100.0	100.0
	Total cases	Total value	Total cases	Total value	Total value
Aggregate cumulative total	693	1,249	1,840	565	156,351

*Value is included in other chemicals

Notes: Values are in millions of U.S. dollars: at current prices for amounts invested and paid; and at constant 1975 prices for value added. Totals may not reconcile due to rounding.

Sources: Information on transfers from Korea Ministry of Finance; that on value added from Bank of Korea, *National Income in Korea*, various years.

TABLE 6.3
Indicators of Technology Inflows, Human Capital, and R & D for Five Semi-industrial Economies

Item	*Year or period*	*Argentina*	*Brazil*	*India*	*Korea*	*Mexico*
Stock of direct foreign investment	1967	10.4	4.0	3.0	1.7	7.3
as a percentage of gross domestic product	1977-79	4.7	6.4	2.1	3.2	5.6
Payments for disembodied technology	1970-71	–	0.20	–	0.04	–
as a percentage of gross national product	1977-79	–	0.33	–	0.17	0.23
Imports of capital goods	1965	5.3	4.6	10.3	13.0	14.5
as a percentage of gross domestic investment	1977-79	8.6	8.4	5.6	27.2	11.8
Postsecondary students abroad	1970	1.0	1.0	1.0	2.0	1.0
as a percentage of all postsecondary students	1975-77	0.3	0.7	0.3	1.7	1.0
Secondary students as a percentage	1965	–	–	29.0	29.0	17.0
of secondary age population	1978	46.0	17.0	30.0	68.0	37.0
Postsecondary students as a percentage	1965	–	–	4.0	5.0	3.0
of eligible postsecondary age population	1978	18.0	10.0	9.0	9.0	9.0
Engineering students as a percentage of total postsecondary age population	1978	14.0	12.0	–	26.0	14.0
Scientists and engineers in thousands	Late 1960s	12.8	5.6	1.9	6.9	6.6
per million of population	Late 1970s	16.5	5.9	3.0	22.0	6.9
Scientists and engineers in R & D	1974	323	75	58	–	101
per million of population	1976	311	–	46	325	–
	1978	313	208	–	398	–

R & D expenditures as a percentage of gross national product	1973	0.3	0.4	0.4	0.3	0.2
	1978	0.4	0.6	0.6	0.7	–

– Not available.

Source: Project files summarizing data from diverse international sources, World Bank Research Project Ref. No. 672-48. Details are available on request from the authors.

such that by the late 1970s South Korea apparently had by far the highest percentage of scientists and engineers in the population of the five countries. It likewise appears to have had proportionately more scientists and engineers engaged in R & D and to have spent proportionately more on R & D. The well-known problems of measurement and interpretation of R & D statistics dictate the need for caution not to read too much into these last comparisons. We nonetheless believe that they are indicative of Korea's relative ability to undertake technological efforts related to assimilation and adaptation.

The Acquisition of Technological Capability

DFI and the modes of disembodied technology imports discussed above are formal means of technology transfer, as is the import of capital goods, which is typically accompanied by disembodied technology in the form of manuals and training. There also are other, informal modes of transfer that span a wide range of possibilities. Evidence about their importance is difficult to obtain, but there is information. Some of it comes from a survey of 113 exporting firms by Pursell and Rhee in 1976.[12] The sample was meant to be representative of all exporting firms in the Republic of Korea.

The firms were asked about the sources of the basic production or process technologies they then used. Domestic sources were considered to be important slightly more often than were foreign sources. For domestic and foreign sources taken jointly, the sources of technology most frequently cited were buyers of output and suppliers of equipment or materials. Next most important were employees with previous experience working in firms overseas – many as a result of training under turnkey and similar arrangements – and in South Korean establishments. Indeed, the transfer of labor among firms was more important than contacts with suppliers alone or with buyers alone.*

*The importance of labor transfer as a source of technology reflects high labor mobility. Depending on the industry, between 33 and 51 percent of the production workers recruited to individual firms in 1975 had previous experience in the job assigned to them.

Formal mechanisms of licensing and technical assistance, of only modest importance overall, were the most important modes for transferring technologies from abroad. Even so, they were considered to be important only one-third as often as foreign sources were indicated. In turn, foreign buyers contributed informal transfers of technology, frequently as a result of periodic visits to inspect production facilities or of ongoing programs to control and improve quality. Through such things as suggesting changes in individual elements of the production process and improvements in the organization of production in the plant and in management techniques more generally, buyers helped many exporters achieve greater efficiency and lower costs. There can be no doubt that the transfer of know-how from export buyers has been a contributor to minor process innovations of the sort that sequentially lead to gradual improvements, the cumulative effect of which can be great.

For many industries it is important to distinguish between the mastery of production processes and the ability to design products that either conform to the structure of or anticipate changes in demand. South Korean exporters, almost across the board, relied heavily on foreign buyers for product design technology, far more so than for process technology. Foreign buyers contributed to product innovation through the influence they exercised on the characteristics of exported products. Nearly three-quarters of the sampled firms stated that they either modified the characteristics of their product to accommodate buyers' requests or produced in direct accord with buyers' specifications. Nonetheless, the majority of firms produced only some of their exports directly to buyers' specifications. Those most often influenced were product design and styling, followed by packaging, basic technical specifications, and minor technical specifications.

It is not surprising that South Korean exporters have relied extensively on foreign sources for product innovation. This is inevitable when technology is transferred to start new lines of production that serve export markets from their inception – as in much of local shipbuilding production, for example. But reliance can also develop if production has first been established to serve the local market, with exports following later, as often occurred in South Korean experience. Here mastery of technology is in the first instance often confined to achieving rudimentary standards of product design. These may suffice to gain entry into export markets, but continued growth of exports sooner or later requires that product standards be upgraded. Moreover, successful penetration of export markets frequently requires that product specifications be tailored to the different demands of individual markets. Until some experience has been gained in producing to meet differentiated demands, it undoubtedly is most cost-effective, and it may even be necessary, to rely on export buyers for product design technology. Not to be neglected in this regard is that production for export provides a potent means of acquiring product design technology through learning-by-doing, which spills over to product development in local markets as well.

The results of the survey of exporters clearly indicate that the acquisition of technological capability by South Korean industry in basic production processes had progressed further than in product design, at least in relation to product standards in developed country export markets. In addition, given the high frequency with which domestic sources of process technology were said to be important, the results attest to considerable local mastery of basic production technology. Much of what was considered by the respondents to have come from domestic sources consisted of technology originally developed overseas, subsequently transferred or brought to South Korea, and then effectively assimilated and sometimes adapted by indigenous industry. Some of this technology, particularly in the traditional export sectors, was part of the inheritance from the colonial past.

The basic production technology for nonsynthetic textile yarn and fabric is an obvious case: among today's leading textile exporters are several that were established before independence, and some senior managers and technicians who gained their initial experience during the colonial period were still active in the 1970s. Plywood – also an important export, particularly during the 1960s – offers another example: the first plant was constructed in 1935. Plywood is equally an example of an industry that benefited from technical assistance provided under the U.S. military's program of local procurement during the 1950s. Nonetheless, when queried about the sources of technologies in use today, producers of both textiles and plywood overwhelmingly indicated local sources.

The distinction between domestic and foreign sources thus has little to do with where the technology was invented. It has far more to do with the importance of the assimilation and adaptation of technology by local producers and of the diffusion of technology through formal and informal contacts and through labor transfers among domestic firms.[13] Further evidence of the importance of diffusion from domestic sources was found in the sizable number of exporting firms that indicated direct knowledge of diffusion to other firms of technologies they had introduced into the country.

In industries for which process technology is not product specific, mastery has frequently led to the copying of foreign products as a means of enlarging technological capacity. The mechanical engineering industries, among others, afford many examples; such processes as machining and casting, once learned through producing one item, can easily be applied in the production of other items. One case that has been closely studied is textile machinery, particularly semiautomatic looms for weaving fabric.[14] In this as in some other cases South Korean manufacturers have not only been able to produce a capital good that meets world standards, albeit for an older vintage; they have, in addition, adapted the product design to make it more appropriate to local circumstances. (The adapted semiautomatic looms fall between ordinary semiautomatic and fully automatic looms in terms of the labor intensity of the weaving technology embodied.)

In other industries in which technology is more product specific, such as chemicals, mastery of the underlying principles has enabled greater local participation in the technological effort associated with the subsequent establishment of closely allied lines of production. Recognition of the importance of local technological learning also is central to understanding how technologies initially introduced in South Korea only very recently – within the past five to ten years – are now considered, in relation to subsequent undertakings in the same lines of production, to have come from local sources.

Detailed information about South Korea's technological capability in particular industries is scanty. Still less information is available about the processes by which its capability has been acquired. In preparing this chapter, Kim undertook an extensive survey of the documentary information available from the records of various government and financial institutions, the objective being to determine whether "technological histories" for specific industries could be compiled from existing project files. He also surveyed existing histories of subsectors and firms. The results were, to say the least, disappointing. It appears that access to historical records is not publicly available, and intensive firm-level interviews provide the only feasible ways to compile even sketchy technological histories. Though disappointed by this finding, we nonetheless feel that there is considerable value in knowing, with some confidence, that there is no feasible alternative.

Detailed historical evidence helps in understanding the evolution in the direction of South Korea's technological development. The rapid growth and increasing diversification of exports of all kinds give the most compelling evidence of the country's acquisition of technological capability over time. But this evidence is indirect and is therefore apt to be misleading in certain important respects. Thus it is important to know that exports have not depended crucially on DFI or other forms of international subcontracting.[15] The survey of exporters clarifies that industrialization up to the late 1970s was based largely on proficiency in production engineering, more specifically on mastery of production processes as opposed to design adaptation. In turn, the extent and nature of the acquisition of investment capability did not become readily apparent until the late 1970s, when South Korea began to experience rapid growth in its exports of elements of technology. But detailed case histories make it clear that a process of increasing involvement in project execution had become entrenched in local industry much earlier.

Amsden and Kim have obtained histories of several turnkey plant exporters in their ongoing research on South Korea's acquisition of technological capability.[16] Two distinct patterns of technology assimilation are apparent. One of them, the apprentice pattern, is well illustrated in the following examples.

- *A chemical company that produces soda ash and other products*. Its initial soda ash plant was built on a turnkey basis by Japanese firms with Japanese technology. Participation in the plant's construction enabled the firm's engineers to do subsequent expansions without foreign assistance. Experience gained through operating the plant and a subsidiary that produces white cement, together with design capabilities accumulated through expansion projects and the establishment of other related plants, gave the firm turnkey capabilities to export a white cement plant.
- *An independent chemical engineering company with three chemical manufacturing affiliates and one construction subsidiary*. To gain engineering experience and exploit the local market

for chemical detergents, the firm imported foreign technology to establish a nonbiodegradable chemical detergent plant. The plant was of West German design, with basic technology from a U.S. company; most of its equipment was imported. The firm's engineers took part in all stages of project execution and benefited from intensive training by the U.S. company. The technological capability so obtained later enabled the firm to build a biodegradable detergent plant without foreign technical assistance. Sixty percent of the equipment installed in this plant was made in South Korea. Experience in these projects and in numerous other engineering projects provided the capabilities to export an unsaturated polyester resin plant to Saudi Arabia. Except for basic engineering of the core processes, the firm undertook all stages of project execution.

Many other industries either are known or appear to have followed the apprentice pattern. For example, Kim, in his supplementary survey for this chapter, obtained sketchy information sufficient to confirm that this pattern prevailed in cement and synthetic fibers (specifically, nylon and polyester).[17] The cement industry is a particularly interesting case because of the continuation of selective foreign participation in basic and detailed engineering, plant start-up, and assistance to quality control in recent projects. Continued reliance on foreigners seems to be linked to efforts to remain close to the global frontier in process technology.

Under the apprentice pattern, the first plant in the industry was typically built on a turnkey basis with indigenous involvement "limited" to assimilating as much of the production *and* investment capability as was practical. The development of investment capabilities began with the participation of South Korean engineers – often more as observers than as anything else – in the initial project execution and continued through experience gained in production and plant expansion projects. Construction of the second and subsequent plants followed quickly, with local engineers and technicians assuming a rapidly expanding role in project design and execution. Indigenous involvement in project implementation expanded through concerted effort to assimilate the know-how involved in project design and execution. The process was effectively one of highly selective and experience-centered import substitution, in which successively more complicated aspects of investment capability were acquired and put into practice. The result was a growing capability in all elements of investment activity.

The rapid accumulation of investment capability through apprenticeship is well illustrated by the expansion path of South Korea's integrated steel mill, which began operation in 1973. The first phase, a turnkey operation by foreigners, created annual capacity of 1 million tons. By 1981 capacity had been increased to 8.5 million tons in three expansion phases that were increasingly under the direction of South Koreans. The progressive substitution of local for foreign investment capability is indicated by the fall in the cost of foreign project engineering per ton of incremental capacity: U.S. $6.13, U.S. $3.81, U.S. $2.42, and U.S. $0.13 respectively in the successive phases. There was also a progressive increase in the share of locally supplied capital goods. Initially 12 percent, it rose to almost 50 percent in the fourth phase. Increased proficiency in investment is further reflected in the shorter times needed to execute successive expansions, even though each phase involved a larger capacity increment: 38 months for the initial, one-million-ton phase, falling to 27 months for the fourth, three-million-ton phase. The time required from start-up to achieve rated capacity also fell in each successive phase. Successful apprenticeship involved large investments in training both at home and abroad. Through 1983 the company had sent 1,850 engineers and technicians for training overseas, much of it involving on-the-job experience.

The other pattern of technological development that appears often is the imitator pattern. It is illustrated in the histories of two other turnkey plant exporters.

- *A local manufacturer of steel pipe, with a subsidiary that manufactures steel-pipe-making equipment.* After the Korean War the firm developed a primitive steel pipe manufacturing line, using automotive parts and other war surplus equipment, based on observation of an imported steel-pipe-manufacturing line used by a local bicycle manufacturer. It later developed and improved six separate lines for its own use before importing several more-advanced lines from Japan. Engineering capabilities developed over time, and insights from operating the Japanese lines led the firm to develop several modern lines of its own based on them. The firm's first turnkey export (to Kenya) was a renovated line that was about to be scrapped because of economic obsolescence.

A second turnkey export (to Bangladesh) was designed and manufactured by the firm. The third was a more sophisticated line than the one used by the firm itself. The Saudi Arabian importer specified automated galvanizing lines to save labor.

- *The only specialized steel rolling mill manufacturer in South Korea*. Originally a machinery repair shop, the firm developed a simple, inexpensive hot-rolling mill for a client. Basic design ideas came from observing an imported mill used by a local firm. Using experience gained from previous projects and technical information obtained from observing more sophisticated foreign mills operating locally and abroad, the firm developed the capability to manufacture a wide range of rolling mills. It designed and sold 102 mills for domestic use before exporting a turnkey plant to Egypt.

As illustrated by these examples, in the imitator pattern, local firms started with small and rather primitive technologies developed by themselves and gradually upgraded both processes and products through operating experience and using technical information and ideas that came from observing foreign technology. Other capital goods producers known to have followed this pattern include suppliers of machinery for textile weaving and paper manufacturing. The fast pace of South Korea's industrial growth has been an important factor in both this and the apprentice pattern. The short intervals between the construction of successive plants to which it led has greatly facilitated experience-based learning.

One of our interests in compiling technological histories was to get some sense of the relative importance of formal and informal technology transfers. Informal transfers have clearly dominated in the imitator pattern and have for a long time been significant in broadening all exporters' capabilities. A study by Kim indicates that they have also been important in recent innovations by small-scale (less than 100 workers) capital goods producers, many of whom initiated new product lines by imitating foreign equipment – by copying imported models and using information from sales catalogues or from visits to foreign manufacturers.[18] We cannot say in precise quantitative terms how important informal transfers have been more generally. But everything we know from firm interviews and other sources indicates clearly that they have been very important.[19]

EXPORTS IN RELATION TO TECHNOLOGICAL DEVELOPMENT[20]

South Korea first adopted a strategy of export promotion in the expectation that it would accelerate growth by relaxing the foreign exchange constraint and increase efficiency through resource allocation in line with comparative advantage. These expectations were more than fulfilled. Since the strategy's adoption, exports have led the country's economic growth. Exports also led the economy's development in a more fundamental sense: in the establishment of new industries and in the acquisition of added technological capability in existing industries.

In this section we first amplify these interactions between export activity and technological capability in general terms. Next we use South Korea's exports of the elements of technology to illustrate the interactions in specific terms. We then use the information about these exports to draw some conclusions about South Korea's technological capability and its comparative advantage in technological activities.

Exports and Technological Capability

South Korea's initial export success came largely in industries, such as textiles, established during the Japanese colonial period, before 1945. Exceptions, such as wigs, used technologies easily assimilated, given the technological capabilities then existing in closely related areas. Thus the underpinnings of export performance during most of the 1960s can be satisfactorily understood in the conventional paradigm of static comparative advantage. Exports were concentrated in industries in which South Korea either already had or could easily acquire the needed technological capability. Moreover, these industries had factor intensities in line with the country's relative factor endowment. The predominant gains from these exports were the obvious ones, greater capacity use and increased allocative efficiency.

It was in the late 1960s that export activity became important in establishing new industries in which South Korea did not already have technological capability. Two wars and their aftermath had opened a hiatus in the establishment of new industries. This was closed in the early 1960s with the inception of two industries – chemical fertilizer and oil refining – both of which were to serve the domestic market with no expectation of substantial exports. Then, in the late 1960s, some new industries were established to serve both the domestic and export markets.

Several of these new industries were characterized by pronounced economies of scale. Constructing plants at scales sufficient only to meet expected domestic demand would have resulted in production costs well above internationally competitive levels. Thus exports were used to gain the economies of scale needed to realize the potential comparative advantage that South Korea had in such industries. Here, a notable example is the integrated manufacture of basic steel products.

Some other new industries had negligible domestic sales at their inception. A few of these, electronic components being an early example, were created by direct foreign investment or relied on other forms of international subcontracting for technology transfer and market access. The rest, such as large-scale shipbuilding, obtained their technology through licensing and turnkey plant contracts and did not have guaranteed markets to begin with.

Export activity can enlarge technological capability in two ways, by facilitating technology transfer and by stimulating technological effort. Transfers of technology often accompany DFI to establish plants that are designed to produce for export. They may also be an integral part of transactions involving other means of international subcontracting. But there are few such instances in South Korea, since DFI and international subcontracting have not been very important in most exports. Export activity can also lead to transfers after the acquisition of rudimentary technological capability. Though less obvious, these transfers have been very important, as was noted previously.

In turn, though there is little direct evidence about the effect that export activity may have in stimulating technological effort, it undoubtedly enforces and fosters the acquisition of technological capability. Exporting requires the ability to meet product specifications at a competitive price, and the drive to penetrate overseas markets stimulates efforts leading to the gradual upgrading of product quality. These may even be the most important ways in which export activity adds to technological capability. But this cannot be directly inferred from studies showing that export activity has a strong positive effect on factor productivity.[21]

In sum, under South Korea's strategy of export-led industrialization, export activity has been important in exploiting static comparative advantage. It has also been important in dynamically changing comparative advantage through accelerating the broadening and deepening of industrial competence. Export activity made it possible to start new industries much earlier than they could otherwise have been established without sacrificing economies of scale. In turn, for all industries, and for a long time after their inception, export activity added to technological capability, reflected in a wide variety of minor technological changes.* Thus export promotion has been a strategy as much for developing industry as for capitalizing on the industrial competence at each point along the way. This is well illustrated in the recent growth of South Korea's exports of the elements of technology.

Exports of the Elements of Technology

Over the past decade South Korea has experienced impressive growth in its exports of the elements of technology, which might also be called exports of capital goods and related services. These exports comprise all flows that involve the transmission of technological knowledge and the performance of activities that reflect its application in establishing and operating productive systems. The constituent elements were indicated and discussed earlier. Here we need only add some remarks about the character of embodiment activities, which, it may be recalled, are those of forming and maintaining physical capital in accord with given and complete design specifications.

---

*To the degree that improvements in productive efficiency and other forms of technological change derive from experience in production and capacity expansion, export activity must necessarily lead to faster productivity growth if it is associated with greater volumes of production over time. This argument is the basis for Abegglen and Rapp's conclusion that export activity has played a major role in the growth of Japanese industrial productivity. Our argument is somewhat different in alleging that export activity has substantial effects that are typically not associated with production for the domestic market. James C. Abegglen and William V. Rapp, "The Competitive Impact of Japanese Growth," in *Pacific Partnership: United States–Japan Trade*, ed. Jerome B. Cohen (Lexington, Mass.: Lexington Books for the Japan Society, 1972), 19-50.

Some capital goods exports consist simply of the manufacture of machinery and equipment (or parts thereof) and the fabrication of structural elements in accordance with detailed design specifications by the purchaser (or an agent acting on its behalf); that is, they consist simply of embodiment activity. Others of these exports, though made to order, may also include domestic design engineering, the scope of which depends on the completeness of specifications provided by the purchaser. Still others comprise machinery and equipment routinely produced in the country. These last two categories of exports also consist of embodiment activity. And insofar as they do not simply embody foreign designs obtained under license, they incorporate capability in the design of capital goods; that is, they incorporate an implicit element of technical services.

Overseas construction, in turn, often consists simply of embodiment activity, as when labor services are supplied with a complementary flow of management services.* The only analytically important difference between this kind of overseas construction and exports of capital goods made to conform to given, detailed design specifications is in the embodiment activity performed. And even this difference disappears for some kinds of overseas construction. Metalworking, as distinct from construction, is the primary embodiment activity in exports of capital goods. But metalworking can also be an important part of overseas construction, as it is, for example, in the erection of chemical plants.

Like other advanced, semi-industrial economies, the Republic of Korea is a substantial exporter of many of the elements of technology. But among these countries it stands out as having the largest volume of such exports.[22] One reason for this is the government's active encouragement of these exports. Also important is the rise to prominence in the mid-1970s of the South Korean *chaebol* – large, conglomerate business groups centered on diversified trading companies. These firms have been the leading agents in the export of the elements of technology.

Table 6.4 provides summary data on what we will call project-related exports, which include all forms of the exports with which

---

* Exports of the elements of technology do not include the services of migrant labor when these are not associated with any other elements, nor do they include flows that consist simply of financial capital or intermediate inputs.

TABLE 6.4
Licensed Project-related Exports by Kind and Sector
(Cumulative, through the end of 1981)

(In millions of U.S. dollars)

	Manufacturing	*Social overhead*	*Services*	*Total*
Overseas construction	2,055	41,777	121	43,953
Plant exports	472	2,098	–	2,570
Direct investment	67	36	–	103
Disembodied technology:				
Licensing and technical agreements	139	13	14	166
Consulting services	155	72	79	306
Total	2,886	43,996	214	47,096

– Zero or not separately distinguished.

Notes: See the text for definitions of kinds and caveats about double-counting. Totals may not reconcile because of rounding error.

Source: Westphal, Larry E., Yang W. Rhee, Linsu Kim, and Alice Amsden, "Exports of Capital Goods and Related Services," from the Republic of Korea, World Bank Staff Working Paper no. 629 (Washington, D.C.: World Bank, 1984), Table 2.

we are concerned except some types of capital goods. Five kinds of project-related exports are distinguished in the table: overseas construction, plant exports, direct (overseas or foreign) investment, licensing and technical agreements, and consulting services. The data are for licensed exports and were tabulated using information from the ministries responsible for licensing them.* (Licensing is done as part of the process of administering export incentives and applies to all exports in the country.)

Overseas construction refers to contracts for construction projects in which the contracting South Korean firm provided more than the services of migrant labor. More will be said about the content of these exports below. Some but not all capital goods exports are included under plant exports. In local usage, plant exports include complete productive systems (such as manufacturing plants and social overhead facilities) and individual elements of such systems (such as textile machinery and distribution transformers). The elements differ from other capital goods exports in that, to be designated as plant exports, they must be purchased for installation in specific productive systems. Moreover, to qualify as a plant export, the transaction has to exceed a minimum value (U.S. $100,000 during the period covered) and satisfy minimum local content requirements that differ by the kind of plant export.

The distinctions among the five kinds of project-related exports are not clear-cut. Most important, contracts that packaged several activities may have been licensed under the dominant activity alone, with the licensed value reflecting the value of the entire package. We know only that this was not a uniform practice. In addition, the reporting of each type of export by a different administrative agency may result in some double-counting. Where direct investment is involved, there is double-counting insofar as the local equity contribution was not in cash but in kind, as frequently was the case. In short, the figures in Table 6.4 are to be taken as indicative rather than definitive. But our impression is that the misclassification and double-counting in these figures is not great enough to make them useless for identifying the central tendencies in Korea's project-related exports.

---

*Data for licensed exports pertain mainly to the value of valid contracts made. Given the lag between contract negotiation and project implementation, the data somewhat overstate the values of actual flows of goods and services.

Overseas Construction

South Korea got into the business of project-related exports through overseas construction, which remains far and away the largest component of these exports. The cumulative value of licensed overseas construction at the end of 1981 was $44 billion, compared with $47 billion for all project-related exports. The composition of contracts for overseas construction exhibits considerable diversity in the size and type of project.

Much of the initial competence in construction activity came from learning-by-doing gained through contracts under U.S. military procurement. Overseas construction began in 1966 and was for some time concentrated in Southeast Asia, where it largely served U.S. military procurement in Guam and Vietnam. The experience gained serving the U.S. military outside South Korea added further capability, specifically in construction work overseas. Thus the stage was set for the dramatic and sustained increase in these exports in 1975, when South Korean contractors began to take part in the Middle East's building boom. Roughly 90 percent of the cumulative value of all contracts has been for work in the Middle East.

Overseas construction appears to consist largely of embodiment activity, though it also includes related organizational and managerial services. Precise information is lacking on the incorporation of technical services related to project design. But the design engineering was South Korean on smaller projects involving buildings – such as schools, warehouses, office structures, apartment blocks, and the like – and simple infrastructure. In turn, the design engineering needed for some large projects, such as those involving large-scale industrial plants, was provided under separate contracts with non-Korean firms.

Plant Exports

South Korea's plant exports began as an adjunct of its overseas construction activity, and further impetus came from the government's promotion of the capital goods sector in the 1970s. Though moving at a much slower pace than the government first hoped, the capital goods sector expanded its capacity quite rapidly, diversifying into new areas. Exports of capital goods in current prices grew at the rate of 60 percent a year during the 1970s, compared with 35 percent for total commodity exports.

Plant exports first reached substantial proportions in 1977. From then onward they may have accounted for as much as one-third of the country's exports of capital goods, with most of the remainder comprising ships and railway vehicles. The cumulative value of the 276 plant exports licensed by the end of 1981 was nearly $2.6 billion, or equal to about 6 percent of the value of licensed overseas construction. Licenses ranged from a U.S. $100,000 contract for power transformers in Indonesia to a $209 million contract for a cement mill in Saudi Arabia.

The provision of a wide variety of manufacturing plants and their elements constituted nearly one-fifth of the value of licensed projects, and more than one-third of the number. The difference in shares by value and by number reflects the small size of manufacturing projects, considerably less than half the average value for social overhead projects. The most important markets for South Korea's manufacturing projects were the Middle East, which accounted for almost 60 percent of the total licensed value, followed by developing countries in Africa and Asia, with roughly 20 percent in each region.

Social overhead projects made up the lion's share of plant exports. They were far more numerous and had an average licensed value of $12.1 million. Very large projects for offshore drilling and coastal facilities were responsible for more than two-thirds of their licensed value. Other projects were related to desalination plants, power generation and transmission facilities, onshore structures, communication facilities, and water treatment plants. Many social overhead projects appear to have been construction related in that they included on-site construction or erection and installation as important elements. The Middle East initially was South Korea's largest market for social overhead projects, but more recently, Organization for Economic Cooperation and Development (OECD) countries have overtaken it. Other developing countries, in Asia and to much less extent in Africa, formed the remainder of the market.

Amsden and Kim have tabulated the elements of 128 plant exports licensed in 1980 and 1981, exports that accounted for nearly half the value of plant exports licensed through 1981.[23] Equipment (including fabricated structural elements) alone was provided in 43 percent of the cases, accounting for 23 percent of the value of the complete sample; equipment and design were provided in 31 percent of the cases, accounting for 35 percent of the sample's value. Fifteen percent of the cases, accounting for 41 percent of the value, were turnkey projects. The remaining 11 percent of the projects, accounting for about 1 percent of the value, combined several elements including equipment but were not full turnkey projects. Both manufacturing and social overhead projects were included in all of these categories.

The absence of design services from some of the manufacturing projects does not mean that the contractor simply provided embodiment services in these cases. On the contrary, most of these projects appear to have supplied South Korean-designed equipment. The absence of design services simply indicates that the designs were standard South Korean designs rather than special ones tailored to the project being served. But circumstances appear to have been different in the social overhead projects: contractors who provided equipment alone probably undertook only the embodiment activities. In turn, design in social overhead projects typically seems to refer to detailed engineering following basic engineering designs provided by the project sponsor. In manufacturing projects, design often appears to have included basic engineering as well.

The driving force behind South Korea's plant exports for social overhead projects was its rapidly advancing comparative advantage in embodiment activity, particularly in large-scale construction projects and capital goods production. The underlying technological capability appears to reside primarily in the organization, management, and execution of construction activity and in the production engineering aspects of metalworking. Additional forces seem to have been at work in the exports of manufacturing projects. Not only did many of these appear to include more sophisticated technical services, such as basic design engineering, but many also appear to have transferred process technologies that incorporated South Korean adaptations. Moreover, several turnkey manufacturing projects called for considerable ability in the organization and management of complex overseas undertakings. In this respect the projects reflected a higher degree of entrepreneurial ability than was probably characteristic of most social overhead projects. Thus the technological capability underlying the manufacturing projects was more extensive in that a much wider range of elements was involved.

Direct Investment

A steady flow of outward investment from South Korea started in about 1967, with a few cases in the preceding years. By the end of 1981 the cumulative licensed outflow of direct investment was U.S. $323 million, of which U.S. $103 million was in manufacturing or social overhead sectors. Thirty-four manufacturing ventures were licensed by the end of 1981. The smallest was for U.S. $25,000 in a printing firm in Japan; the largest was for U.S. $25.7 million in a cement mill in Malaysia. About three-quarters of this investment was in developing countries in Asia; most of the rest was in Africa and the Middle East.

There was a considerable overlap between overseas manufacturing ventures and manufacturing plant exports. Indeed, in some cases an equity position is either known or appears to have been taken to facilitate the export of a plant. But some investments were motivated more by the desire for access to natural resources, albeit in a processed form, than by anything else. Another factor motivating some joint-venture investments, particularly those in the OECD countries, was the desire to gain access to foreign technology through collaboration with foreign companies.

Disembodied Technology Exports

In South Korean usage, licensing and technical agreements are distinguished from technical consulting services. But the dividing line between the two categories is not clear-cut. The only real distinction we know of is that no contracts involving royalty payments over time (as opposed to one-time lump-sum payments) are licensed under consulting services.

No licensing or technical agreements were recorded before 1978, when two contracts were licensed – we do not know their value. Of the 39 contracts worth U.S. $166 million licensed during 1979-81, 36 were with firms in developing countries, mainly in the Middle East. Most licensing and technical agreements provided general technical assistance in production engineering and maintenance, many over several years. The sectors served were mainly in manufacturing, with a rather wide dispersion among these sectors. And some of the contracts were to plants in whose establishment South Korean firms had participated through overseas construction or plant exports.

Exports of consulting services were far more numerous than those of licensing and technical agreements. The cumulative value of the 324 contracts licensed from 1973 through 1984 was U.S. $306 million, with more than three-quarters of the contracts (by number and value) having been licensed in 1979-81. Most comprised technical services accomplished by the dispatch of personnel to oversee construction and plant erection, install machinery and equipment, inspect structures and equipment, troubleshoot, train labor, and the like. The value of licensed contracts was split about evenly between manufacturing and social overhead sectors. Many consulting services, particularly in the social overhead sectors, appear to have been related to construction exports. Firms specializing in engineering appear to have been responsible for most exports of both licensing and technical agreements and consulting services – not only exports related to project execution, but also those involving production engineering. These exports thus reflect the growing capability of South Korea's recently established engineering firms.[24]

Revealed and Dynamic Comparative Advantage

We have discussed five kinds of project-related exports – overseas construction, plant exports, direct investment, licensing and technical agreements, and consulting services. These exports and those of non-project-related capital goods constitute the exports of elements of technology. Overseas construction and plant exports for social overhead projects have predominated. They accounted for more than 95 percent of the cumulative value of licensed project-related exports at the end of 1981 and around half to three-quarters of total exports of capital goods and other elements of technology during 1977-81. The bulk of this export activity appears to have been performed in accordance with detailed specifications provided by the purchaser. We thus conclude that South Korea's revealed comparative advantage in exports of the elements of technology is in project execution, mostly in embodiment activities in the form of construction and metalworking (including erection and fabrication).*

---

*We are constrained to assess revealed comparative advantage by the composition of Korea's exports. But we doubt that the use of more sophisticated methods would change the conclusion.

These embodiment activities are moderately intensive in human capital. They require reasonably skilled workers and technicians as well as competent engineers, of whom the country has a comparative abundance. But more than this, they require the ability to organize and manage undertakings that are often complex, even in subcontracts. Here South Korean firms appear to have an advantage that permits them, for example, to complete projects in far less time than is considered average or normal. Precise information about this is lacking but anecdotes abound. Moreover, much of the marketing of these exports is done by South Korean firms acting without foreign agents. This is one area where South Korean know-how relating to transactions is second to none.

In most of its exports of the elements of technology South Korea is not exporting newly created technological knowledge. Instead, these exports largely consist of certain project execution activities. In other words, what underlies the country's revealed comparative advantage is its mastery of production engineering in construction and metalworking. It enjoys a cost advantage in these activities due both to its mastery and to its comparatively low wages and salaries – adjusted for skill and productivity differences – for skilled workers, technicians, and managers.

A small part of South Korea's exports nonetheless do appear to transfer idiosyncratic manufacturing technologies created through experience-based adaptive engineering. In turn, technological idiosyncrasies may be an element in the revealed comparative advantage in project execution. That is, the technologies used to carry out these activities might be idiosyncratic. Differences in technology related to organizational modes and procedural methods might, for example, be what underlies the ability of firms to complete projects in record time. This possibility is certainly consistent with our conception of the elements of investment capability.

The technology factor that underlies most of South Korea's exports of the elements of technology is much the same as that which underlies most of its (other) manufactured exports. Insofar as the former exports largely reflect mastery of what might be termed the production engineering aspects of project execution, the rapid growth of both kinds of exports reflects the rapid accumulation of proficiency in production. But the emphasis on proficiency in production should not imply the lack of a design capability. Exports of idiosyncratic manufacturing technology indicate that there is some capability in the design of machinery. And South Korea does export detailed project-engineering services. But these exports are only a small fraction of the total. What is more significant is that South Korea does not appear to possess much capability in basic project engineering.

In discussing the general relationship between exports and technological capability in South Korea, we emphasized two distinct objectives – exploiting the country's existing comparative advantage, and dynamically changing its comparative advantage. There can be little question, especially in the light of the foregoing discussion, that South Korea's exports of elements of technology exploit its existing comparative advantage, in both its endowment of human capital and its mastery of the production engineering aspects of project execution. What, then, of their role in dynamically changing the country's comparative advantage?

Because of their highly specialized nature, many activities of project execution are characterized by extreme economies of scale.[25] South Korean firms could not be internationally competitive in these activities if they served only the domestic market. Export activity has thus made it possible to establish investment capabilities that could not otherwise have been realized without tremendous sacrifice of scale economies. It has also enabled greater continuity over time in the use of capabilities once established, thereby precluding their atrophy owing to infrequent use. Moreover, the accumulation of experience is a critical input in acquiring most of these capabilities.[26] Export activity not only compresses the time for experience to be accumulated; it also affords a wider variety of experience in more diverse circumstances. It can thus be expected to accelerate cost reductions from learning and to deepen existing capabilities.

These benefits appear to be reflected in changes over time in the composition of South Korea's exports of elements of technology toward increasingly more complex and sophisticated activities. Managers of exporting firms have also confirmed, in interviews with Amsden and Kim, that these benefits are realized.[27] But these managers have indicated another benefit as being even more significant: participation in project execution with foreign firms from third countries has been an important vehicle for acquiring additional capabilities and new technologies. These gains have occurred through a process akin to apprenticeship – a process also observed (as previously noted) in South Korean participation in local projects. This broadening of technological capability occurs even in overseas construction. For example, one South Korean firm assimilated a complete system of solar energy technology through its participation with a U.S. firm in a project in Saudi Arabia.

The broadening and deepening of South Korea's industrial competence, particularly its investment capability, appears to be an important motive in the government's promotion of exports of the elements of technology. In the short run, the gains from exploiting existing comparative advantage in these exports are considerable. But in the long run, the gains from further developing the country's technological capability may well be even more considerable.

CONCLUSION

Several things stand out in the Republic of Korea's pattern of technological development. One is the limited extent of reliance on proprietary transfers of technology by means of direct foreign investment and licensing agreements. Among formal transfers of technology, turnkey plants and machinery imports have played by far the greater role. Moreover, in only a few sectors, such as electronics, have exports depended crucially on transactions between related affiliates of multinational corporations or on other forms of international subcontracting. Another prominent feature of South Korean technological development is the apparently tremendous importance of a wide variety of informal transfers that have involved imitation and apprenticeship as well as the use of information obtained in exporting. Also to be included under informal transfers is the expertise that has been obtained as a result of the return of South Koreans from study or work abroad unassociated with formal training, though the importance of this transfer relative to formal training is not known.

Also striking is the selectivity of South Korean technological development and the part played in it by both imports and exports of the elements of technology. South Koreans have acquired a good deal of technological capability, but they have done so in piecemeal fashion as successively more sophisticated capabilities have been acquired and put into practice. The process of acquisition has clearly been one of purposive effort involving a succession of incremental steps, with production capabilities being developed somewhat in advance of investment capabilities. The selectivity of import substitution for the elements of technology has meant continued reliance on imports for at least some elements in almost all industries, but the pattern of imports has continually shifted as local capabilities have replaced foreign ones and as new industries have been developed. In turn, the selectivity of import substitution is complemented in the pattern of exports by specialization among the elements of technology in line with what one would expect to be South Korea's comparative advantage. And there is evidence that the pattern of exports, like the pattern of imports, has been shifting rapidly in response to the country's fast-paced acquisition of technological capability in many areas. What is perhaps most remarkable in this respect is that export activity appears to be an important vehicle for acquiring additional technological capabilities.

South Korea's ability to industrialize without extensive reliance on proprietary transfers of technology is in part explained by the nature of technology and product differentiation in the industries on which its growth has so far crucially depended. Many of these industries – such as plywood, textiles, and apparel – use relatively mature technologies; in such cases, mastery of well-established and conventional methods, embodied in equipment readily available from foreign suppliers, is sufficient to permit efficient production. The products of many of these industries are either quite highly standardized (plywood, for example) or differentiated in technologically minor respects and not greatly dependent on brand recognition for purchaser acceptance (textiles and apparel, for example). Thus, in most of the industries that have been intensively developed, few advantages are to be gained from licensing or direct foreign investment as far as technology acquisition and overseas marketing are concerned.

Nonetheless, exceptions exist, most notably in the chemical industry, where South Korea has had to rely extensively on direct foreign investment to establish and expand production, no doubt because of the reluctance of the technology suppliers to transfer technology via other modes. But in other industries where technology is also proprietary, a number of examples attest to the fact that South Korean industry has managed to initiate, and in most cases to operate successfully, a variety of high-technology industrial activities by means of licensing and turnkey arrangements. To cite two cases: arrangements of this kind were used to acquire the most modern shipbuilding technology in the world, and to incorporate the most recent technological advances in the integrated steel mill.

More generally, the country's recent experience in promoting technologically sophisticated industries indicates that their development may involve greater reliance on licensing as a way of acquiring technology. It is also probable that the shift to new industries may imply greater dependence on direct foreign investment. Indeed, greater dependence on such proprietary transfers of technology is observed starting in the latter half of the 1970s. But for the period in which semi-industrial status was achieved – that is, from the mid-1960s to the mid-1970s – there can be no doubt about the overwhelming role of indigenous effort in using capital goods imports and informal transfers of technology to acquire technological capability.

What is unique about the South Korean experience is not the importance of indigenous effort to assimilate technology. As we indicated at the outset, research on all semi-industrial economies shows that such effort is crucial for the acquisition of technological capability. What is unique about the South Korean experience is the speed and effectiveness of acquisition and the interplay between technological development and trade in the elements of technology. Near the beginning of the chapter, we outlined a framework for understanding and analyzing that interplay. The framework, like everything known about technological development, suggests the importance of paying attention to fine distinctions among technological capabilities when trying to understand the role of experience-based efforts in the cumulative buildup of technological capability. Because of space limitations, our discussion of South Korean technological development in later sections glossed over some of these distinctions.

Although we have enough details to conclude that such distinctions are fundamental, we lack a sufficient base of knowledge to generalize from South Korea's experience about many key aspects of formulating a strategy for technological development. We would emphasize four areas where more knowledge needs to be sought in order to understand the sources of South Korean success. One is the positive relationship between education and technological development – the details of this relationship remain to be uncovered. Another is the difference between competition in domestic and export markets in stimulating technological effort. We suspect but cannot yet demonstrate that an export-promoting policy regime provides a far more effective stimulus than either an inward-looking regime or an outward-looking one that simply aims at international competitiveness.

The third area is the role of government policy interventions in fostering South Korea's export-led industrialization and in stimulating technological development more generally. We know enough to conclude that the government's role cannot be characterized or judged in simple terms. We have not provided much discussion of the government's role because we wanted to focus on describing the main features of the country's technological development. Another reason is that more needs to be known before many conclusions can be reached. The last area for study is the importance of "initial conditions" in the early 1960s. Does rapid, sustained technological development in industry require the prior attainment of some minimal level of industrial competence built up through a process that is either much less rapidly paced or far more dependent on foreign production capability obtained through direct foreign investment or the use of expatriate manpower? The question remains wide open. Combining further study in these areas with the findings of complementary research on other semi-industrial economies may move us closer to answering important questions about the formulation of strategies for technological development.

NOTES

1. This section builds on Carl J. Dahlman and Larry E. Westphal, "Technological Effort in Industrial Development – An Interpretive Survey of Recent Research," in *The Economics of New Technology in Developing Countries*, ed. Frances Stewart and Jeffrey James (Boulder, Colo.: Westview Press, 1982), 105-37; and Dahlman and Westphal, "The Transfer of Technology: Issues in the Acquisition of Technological Capability by Developing Countries," *Finance & Development* 20 (December 1983), 6-9. The former paper uses *mastery* in much the same sense that *capability* is used here. Nelson's overview of issues in technological development also provides a pertinent perspective from which to understand the framework expounded in this section. Richard Nelson, "Research on Productivity Growth and Productivity Differences: Dead Ends and New Departures," *Journal of Economic Literature* 19 (September 1981), 1029-64.

2. For supporting argument and a survey of the relevant empirical evidence on these points, see Dahlman and Westphal, "Technological Effort in Industrial Development."

3. Adaptations have been observed to take place through changes that stretch the capacity of existing plants, break bottlenecks in particular processes, improve the use of by-products, adjust to new input sources, alter the product mix, and introduce a wide variety of incremental improvements in processes and product designs. For summaries of the most extensive program of case history research to date, see Jorge Katz, "Technological Change, Economic Development and Intra and Extra Regional Relations in Latin America," IDB/ECLA/UNDP Research Program on Scientific and Technological Development in Latin America, Working Paper no. 30 (Buenos Aires: United Nations Economic Commission for Latin America, 1978); Katz, "Domestic Technology Generation in LDCs: A Review of Research Findings," IDB/ECLA/UNDP Research Program on Scientific and Technological Development in Latin America, Working Paper no. 35 (Buenos Aires: United Nations Economic Commission for Latin America, 1980); Katz, "Technological Change in the Latin American Metalworking Industry: Results of a Program of Case Studies," IDB/ECLA/UNDP Research Program on Scientific and Technological Development in Latin America, Working Paper no. 51 (Buenos Aires: United Nations Economic Commission for Latin America, 1982); and Simon Teitel, "Creation of Technology within Latin America," *Annals of the American Academy of Political and Social Science*, 458 (November 1981), 136-50.

4. Rosenberg forcefully demonstrates this from the perspective of economic history. Nathan Rosenberg, *Perspectives on Technology* (Cambridge: Cambridge University Press, 1976).

5. Sang Chul Suh, *Growth and Structural Changes in the Korean Economy, 1910-1940* (Cambridge, Mass.: Harvard University Press for the Council on East Asian Studies, Harvard University, 1978), p. 153.

6. Edward S. Mason et al., *The Economic and Social Modernization of the Republic of Korea*, Studies in the Modernization of the Republic of Korea, 1945-1975, Harvard East Asian Monographs, vol. 92 (Cambridge, Mass.: Harvard University Press for the Council on East Asian Studies, Harvard University, 1980), p. 449.

7. For details, see George M. McCune, *Korea Today* (Cambridge, Mass.: Harvard University Press, 1950), chaps. 3 and 8.

8. For details about American aid, see Anne O. Krueger, *The Developmental Role of the Foreign Sector and Aid*, Studies in the Modernization of the Republic of Korea, 1945-75, Harvard East Asian Monographs, vol. 87 (Cambridge, Mass.: Harvard University Press for the Council on East Asian Studies, Harvard University, 1979) and Mason et al., *The Economic and Social Modernization of the Republic of Korea.*

9. For information about these reforms, see Larry E. Westphal, "The Republic of Korea's Experience with Export-Led Industrial Development," *World Development* 6 (March 1978), 347-82, and Mason et al., *The Economic and Social Modernization of the Republic of Korea.*

10. Much of this section is based on Larry E. Westphal, Yung W. Rhee, and Garry Pursell, "Korean Industrial Competence: Where It Came from," World Bank Staff Working Paper no. 469 (Washington, D.C.: World Bank, 1981), and Larry E. Westphal, Yung W. Rhee, Linsu Kim, and Alice Amsden, "Exports

of Capital Goods and Related Services from the Republic of Korea," World Bank Staff Working Paper no. 629 (Washington, D.C.: World Bank, 1984).

11. Westphal et al., "Korean Industrial Competence."

12. See Westphal et al., "Korean Industrial Competence," pp. 38 ff.

13. Kim provides a model of the underlying process of technological development and illustrates it with reference to South Korea's electronics industry. Linsu Kim, "Stages of Development of Industrial Technology in a Developing Country: A Model," *Research Policy* 9 (July 1980), 254-77.

14. Yung W. Rhee and Larry E. Westphal, "A Micro Econometric Investigation of Choice of Technology," *Journal of Development Economics* 4 (September 1977), 205-37.

15. On international subcontracting, see Westphal et al., "Korean Industrial Competence," pp. 52 ff.

16. Alice H. Amsden and Linsu Kim, "Korea's Technology Exports and Acquisition of Technological Capability" (Washington, D.C.: Productivity Division, Development Research Department, World Bank, 1982), second draft.

17. Additional technological histories – for petrochemicals, paper products, nylon yarn and cord, and the integrated manufacture of basic steel products – displaying the same pattern appear in Seoul National University, "The Absorption and Diffusion of Imported Technology in Korea" (Seoul: Institute of Economic Research, Seoul National University, 1980), draft, and in John Enos, "The Choice of Techniques vs. the Choice of Beneficiary: What the Third World Chooses," in *The Economics of New Technology in Developing Countries*, ed. Stewart and James, 69-81.

18. Linsu Kim, "Technological Innovations in Korea's Capital Goods Industry: A Micro Analysis," ILO Working Paper WEP 2-22/WP 92 (Geneva: International Labor Office, 1982).

19. Linsu Kim and Youngbae Kim, "Innovation in a Newly Industrializing Country: A Multiple Discriminant Analysis," *Management Science*, forthcoming.

20. Much of this section is taken without further citation from Westphal et al., "Exports of Capital Goods and Related Services."

21. See Bela Balassa, "Development Strategies and Economic Performance: A Comparative Analysis of Eleven Semi-Industrial Economies," in *Development Strategies in Semi-Industrial Economies*, Bela Balassa et al. (Baltimore: Johns Hopkins University Press for the World Bank, 1982), 38-62; Gershon Feder, "On Exports and Economic Growth," *Journal of Development Economics* 12 (February-April 1983), 59-74; and Mieko Nishimizu and Sherman Robinson, "Trade Policies and Productivity Change in Semi-Industrialized Countries," Development Research Department Discussion Paper no. 52 (Washington, D.C.: World Bank, 1983).

22. See Carl J. Dahlman and Francisco Sercovich, "Exports of Technology from Semi-Industrial Economies and Local Technological Development," *Journal of Development Economics* 16 (September-October 1984), 63-99.

23. Amsden and Kim, "Korea's Technology Exports."

24. See Jinjoo Lee, "Development of Engineering Consultancy and Design Capability in Korea," in *Consulting and Engineering Design in Developing*

Countries, ed. Alberto Araoz (Ottawa: International Development Research Center, 1981), 61-78.

25. See Alberto Araoz, "Consulting and Engineering Design Organizations in Developing Countries," in *Consulting and Engineering Design in Developing Countries*, ed. Alberto Araoz (Ottawa: International Development Research Center, 1981), 9-52; Martin Brown, "Engineering of Industrial Projects: Some Reflections on Current Development Center Research," Industry and Technology Occasional Paper no. 11 (Paris: OECD Development Center, 1976), document no. DC/TI(76)4; Martin Brown and Jacques Perrin, with the assistance of Dominique Genet, "Engineering and Industrial Projects: A Survey of Engineering Service Organizations" (Paris: OECD Development Center, 1977), document no. CD/R(77)2; and David J. Teece, *The Multinational Corporation and the Resource Cost of International Technology Transfer* (Cambridge, Mass.: Ballinger, 1976).

26. See Araoz, "Consulting and Engineering Design Organizations"; Brown, "Engineering of Industrial Projects"; Brown and Perrin, "Engineering and Industrial Projects"; Teece, *The Multinational Corporation*.

27. Amsden and Kim, "Korea's Technology Exports."

REFERENCES

Abegglen, James C., and William V. Rapp. "The Competitive Impact of Japanese Growth." In Jerome B. Cohen, ed., *Pacific Partnership: United States–Japan Trade*, pp. 19-50. Lexington, Mass.: Lexington Books for the Japan Society, 1972.

Amsden, Alice H., and Linsu Kim. "Korea's Technology Exports and Acquisition of Technological Capability." Washington, D.C.: Productivity Division, Development Research Department, World Bank, 1982. Second draft.

Araoz, Alberto. "Consulting and Engineering Design Organizations in Developing Countries." In Alberto Araoz, ed., *Consulting and Engineering Design in Developing Countries*, pp. 9-52. Ottawa: International Development Research Center, 1981.

Balassa, Bela. "Development Strategies and Economic Performance: A Comparative Analysis of Eleven Semi-Industrial Economies." In Bela Balassa and Julio Berlinski, Ow Chin Hock, Thomas L. Hutcheson, Kwang-Suk Kim, T. H. Lee, Kwo-shu Liang, Daniel M. Schydlowsky, Zvi Sussman, Augustine H. H. Tan, and Larry E. Westphal, *Development Strategies in Semi-Industrial Economies*, pp. 38-62. Baltimore: Johns Hopkins University Press for the World Bank, 1982.

Brown, Martin. "Engineering of Industrial Projects: Some Reflections on Current Development Center Research." Industry and Technology Occasional Paper no. 11, document no. DC/TI(76)4. Paris: OECD Development Center, 1976.

Brown, Martin, and Jacques Perrin, with the assistance of Dominique Genet. "Engineering and Industrial Projects: A Survey of Engineering Service Organizations." Document no. CD/R(77)2. Paris: OECD Development Center, 1977.

Dahlman, Carl J., and Francisco Sercovich. "Exports of Technology from Semi-Industrial Economies and Local Technological Development." *Journal of Development Economics* 16 (September-October 1984): pp. 63-99.

Dahlman, Carl J., and Larry E. Westphal. "Technological Effort in Industrial Development – An Interpretive Survey of Recent Research." In Frances Stewart and Jeffrey James, eds., *The Economics of New Technology in Developing Countries*, pp. 105-37. Boulder, Colo.: Westview Press, 1982.

——. "The Transfer of Technology: Issues in the Acquisition of Technological Capability by Developing Countries." *Finance & Development* 20 (December 1983): pp. 6-9.

Enos, John. "The Choice of Technique vs. the Choice of Beneficiary: What the Third World Chooses." In Frances Stewart and Jeffrey James, eds., *The Economics of New Technology in Developing Countries*, pp. 69-81. Boulder, Colo.: Westview Press, 1982.

Feder, Gershon. "On Exports and Economic Growth." *Journal of Development Economics* 12 (February/April 1983): pp. 59-74.

Hall, G. R., and R. E. Johnson. "Transfers of United States Aerospace Technology to Japan." In Raymond Vernon, ed., *The Technology Factor in International Trade*, pp. 305-58. New York: National Bureau of Economic Research, distributed by Columbia University Press, 1970.

Hayami, Yujiro, and Vernon W. Ruttan. *Agricultural Development: An International Perspective*. Baltimore: Johns Hopkins Press, 1971.

Katz, Jorge. "Domestic Technology Generation in LDCs: A Review of Research Findings," IDB/ECLA/UNDP Research Program on Scientific and Technological Development in Latin America, Working Paper no. 35. Buenos Aires: United Nations Economic Commission for Latin America, 1980.

——. "Technological Change, Economic Development and Intra and Extra Regional Relations in Latin America." IDB/ECLA/UNDP Research Program on Scientific and Technological Development in Latin America, Working Paper no. 30. Buenos Aires: United Nations Economic Commission for Latin America, 1978.

——. "Technological Change in the Latin American Metalworking Industry: Results of a Program of Case Studies." IDB/ECLA/UNDP Research Program on Scientific and Technological Development in Latin America, Working Paper no. 51. Buenos Aires: United Nations Economic Commission for Latin America, 1982.

Kim, Linsu. "Stages of Development of Industrial Technology in a Developing Country: A Model." *Research Policy* 9 (July 1980): pp. 254-77.

——. "Technological Innovations in Korea's Capital Goods Industry: A Micro Analysis." ILO Working Paper WEP 2-22/WP 92. Geneva: International Labor Office, 1982.

Kim, Linsu, and Youngbae Kim. "Innovation in a Newly Industrializing Country: A Multiple Discriminant Analysis." *Management Science*, forthcoming.

Krueger, Anne O. *The Developmental Role of the Foreign Sector and Aid*. Studies in the Modernization of the Republic of Korea, 1945-75, Harvard East Asian Monographs, vol. 87. Cambridge, Mass.: Harvard University Press for the Council on East Asian Studies, Harvard University, 1979.

Lee, Jinjoo. "Development of Engineering Consultancy and Design Capability in Korea." In Alberto Araoz, ed., *Consulting and Engineering Design in Developing Countries*, pp. 61-78. Ottawa: International Development Research Center, 1981.

McCune, George M. *Korea Today*. Cambridge, Mass.: Harvard University Press, 1950.

Mason, Edward S., Mahn Je Kim, Dwight H. Perkins, Kwang-Suk Kim, and David C. Cole. *The Economic and Social Modernization of the Republic of Korea*. Studies in the Modernization of the Republic of Korea, 1945-75, Harvard East Asian Monographs, vol. 92. Cambridge, Mass.: Harvard University Press for the Council on East Asian Studies, Harvard University, 1980.

Nelson, Richard. "Research on Productivity Growth and Productivity Differences: Dead Ends and New Departures." *Journal of Economic Literature* 19 (September 1981): pp. 1029-64.

Nishimizu, Mieko, and Sherman Robinson. "Trade Policies and Productivity Change in Semi-Industrialized Countries." Development Research Department Discussion Paper no. 52. Washington, D.C.: World Bank, 1983.

Rhee, Yung W., and Larry E. Westphal. "A Micro Econometric Investigation of Choice of Technology." *Journal of Development Economics* 4 (September 1977): pp. 205-37.

Rosenberg, Nathan. *Perspectives on Technology*. Cambridge: Cambridge University Press, 1976.

Salter, W. E. G. *Productivity and Technical Change*. Cambridge: Cambridge University Press, 1960.

Seoul National University. "The Absorption and Diffusion of Imported Technology in Korea." Seoul: Seoul National University, Institute of Economic Research, 1980. Draft.

Suh, Sang Chul. *Growth and Structural Change in the Korean Economy, 1910-1940*. Cambridge, Mass.: Harvard University Press for the Council on East Asian Studies, Harvard University, 1978.

Teece, David J. *The Multinational Corporation and the Resource Cost of International Technology Transfer*. Cambridge, Mass.: Ballinger, 1976.

Teitel, Simon. "Creation of Technology within Latin America." *Annals of the American Academy of Political and Social Science* 458 (November 1981): pp. 136-50.

Westphal, Larry E. "The Republic of Korea's Experience with Export-Led Industrial Development." *World Development* 6 (March 1978): pp. 347-82.

Westphal, Larry E., Yung W. Rhee, Linsu Kim, and Alice Amsden. "Exports of Capital Goods and Related Services from the Republic of Korea." World Bank Staff Working Paper no. 629. Washington, D.C.: World Bank, 1984.

Westphal, Larry E., Yung W. Rhee, and Garry Pursell. "Korean Industrial Competence: Where It Came From." World Bank Staff Working Paper no. 469. Washington, D.C.: World Bank, 1981.

7

Macroeconomic Factors Affecting Japan's Technology Inflows and Outflows: The Postwar Experience

Terutomo Ozawa

INTRODUCTION

Japan's postwar technological progress has been quite impressive by any standard. Japan now leads the world in several areas and in fact has emerged as an active exporter of technology. Although it is still the world's most eager absorber of foreign industrial arts, its overall balance of technology trade has recently turned to a substantial surplus so far as new contracts are concerned.

In 1949, Japan's labor productivity, reflecting an extremely low level of industrial technology, was only a small fraction of the U.S. level: 5 percent in the coal and chemical industries and 10 percent in the rubber industry. Even in the rayon fiber industry, in which Japan had already developed a strong comparative advantage before the war, it was 20 percent. It was, indeed, recognized, not without a sense of resignation, that it would take Japan anywhere from 10 to 30 years just to close the then existing technology gap with the West, not to speak of learning new postwar technologies.[1]

But thanks to a series of favorable events, particularly the procurement boom created by the Korean War, Japan had by the mid-1950s restored its industrial output to the pre-World War II level. In 1956, as a result, the Japanese government openly declared, "We are no longer in the postwar period." Modernization of productive facilities through the use of the latest technologies has been the key feature of Japan's economic miracle.

How, then, did Japan accomplish such technological prowess? One can perhaps readily respond by saying that it has succeeded in closing the huge technology gap by simply borrowing the latest technologies from the West: the larger the technology gap to be filled and the greater the absorptive capacity of a follower country, Japan, the faster the rate of technological advance and industrial expansion in that country. But such a statement, though it may be correct, is obviously too sweeping to explain anything. The process of catching up – and surpassing in some sectors under the very momentum of catching up – is a very complicated one and involves a multitude of factors and their optimal combinations. To begin with, for example, there must be the ability to absorb and master modern industrial knowledge at the enterprise level; a ready supply of advanced technologies; and, above all, a sufficiently effective demand for the products to be manufactured by new technologies, as well as a social mobilization of capital to be invested in large-scale productive facilities, a carrier of modern technology. In other words, both internal faculty (such as the firm's motivation and ability to master technology) and external stimuli (such as the availability of technology to be commercialized and the existence of effective demand) need to exist if a successful technological absorption on a large scale, as occurred in Japan, is to take place.

A study of technological progress can – and should – be carried out at different levels: macro (that of the economy), micro (that of the industry and firm), and micro-micro (those within the firm). This chapter focuses only on some key macroeconomic (both intertemporal and cross-sectional) factors that have facilitated and influenced Japan's postwar technological progress. I will first emphasize the fundamental long-term historical force behind the postwar experience and the crucial role of the World War II period in giving Japan an "appropriate" pose in its century-old drive to catch up with the West. With these historical forces as a backdrop I will then examine each postwar decade, that is, the 1950s, the 1960s, and the 1970s, in chronological order, for in each decade Japan's technological effort was carried out in a different macroeconomic environment.

TECHNOLOGY INFLOWS

Although it is only in recent years that Japan's innovative capacity has been recognized by the world community, it needs to be looked at as a continuous process dating back many centuries. Japan's cultural heritage is such that it firmly believes in its ability to learn from outside *selectively* without disrupting its own fabric of tradition. Japan has always been – and is – ready to absorb anything superior by adapting it to its indigenous environment.

The Legacy of Many Centuries

Japan's first experience with adopting elements of a foreign civilization on a massive scale was in the seventh century, when it borrowed systematically a wide range of social and even religious institutions from the T'ang dynasty of China. Buddhism and Confucianism were adopted as effective instruments to unify a nation still dominated by local clans. A number of significant industrial arts, such as architecture, civil engineering, weaving, pottery, lacquerware making, mining, metallurgy, and farming, were also brought back home by the numerous Japanese missions called *Kentoshi* ("those dispatched to T'ang"). This first wave of cultural assimilation continued until the T'ang dynasty declined in power in the late ninth century. Later on, in the twelfth century and, again, in the fifteenth century Japan had active contacts with China, in the reign of the Sung and the Ming dynasties respectively.[2]

Japan's direct contact with the West took place for the first time in the mid-sixteenth century, when shipwrecked Portuguese seamen were washed ashore at the southern tip of Japan's archipelago. How quickly Japanese *daimyos* learned the use of muskets from them and reproduced the firearms in large quantities is a well-known story. Clocks and clockwork were also soon introduced by the Portuguese: "The Japanese, skilled as they were in the metalwork, did not have much trouble copying the mechanisms."[3]

To avoid the disruptive effect of Christian evangelism on the absolute power of the Shogunate, Japan limited its contact with the West by adopting a policy of isolation, which lasted about 250 years starting in 1640. During this long period of seclusion, however, Japan kept track of developments abroad via an islet called Dejima in the province of Nagasaki, where the Dutch, English, and Chinese were permitted to come and stay to trade. Japan thus learned, though on a limited basis, modern Western sciences such as astronomy, mathematics, and human anatomy.

When Japan finally came out from this long period of isolationism and embarked on an ambitious program to modernize its socioeconomic institutions by adopting both material and social technologies from the West, it had by then already accumulated many centuries' experience of selecting and assimilating superior foreign cultures into its own social fabric. The national slogan adopted after 1868 was, "Western technology and the Japanese ethos."[4] Western technology was merely a means to enhance Japan's own strength.

Thus Japan's assimilative effort runs through its entire existence. Its history has long been characterized by the ability to learn selectively and to indigenize things alien. In this connection it is interesting to observe a close similarity between Europe and Japan: both exhibited a high propensity to learn from outside.

> Perhaps European civilization could not have progressed so rapidly had it not possessed a remarkable faculty for assimilation – from Islam, from China, and from India. No other civilization seems to have been so widespread in its roots, so eclectic in its borrowings, so ready to embrace the exotic. Most have tended (like the Chinese) to be strongly xenophobic, and to have resisted confession of inferiority in any aspect, technological or otherwise. Europe would yield nothing of the preeminence of its religion and but little of its philosophy, but in processes of manufacture and in natural science it readily adopted whatever seemed useful and expedient.[5]

In this similar cultural trait may lie an important clue as to why Japan has succeeded in becoming the only non-Western nation so far to join the ranks of the industrial powers. A high receptivity toward alien cultures and the actual historical experience of assimilating aspects of them, especially industrial knowledge, underlie Japan's postwar technological achievement.

In addition, emphasis on learning as a cultural trait also formed the basis for human resource development. Even before the Meiji Restoration of 1868, which marked the beginning of Japan's modern era, literacy rates (the ability to read public notices, if not literary works) among the key economic decision makers were surprisingly high.

> On the average, close to 40 percent of the adult male population was already somewhat literate. . . . It is estimated that literacy among samurai and the nobility was virtually 100 percent; but literacy rates also ran as high as 70 to 80 percent for the large city merchants, 50 to 60 percent for the artisans and small town traders, and at least 20 percent even for the peasantry.[6]

Indeed, one of the policy priorities of the Meiji government for modernization was to introduce a nationwide education system under which all children from 6 through 13 years of age were required to attend school. Human resource development through education, particularly designed to build individuals' capacities and skills needed for modern industrial activities, has ever since been a continuous national commitment. It constituted a vital foundation for technological assimilation and development.

Wartime Technological Autarky: An "Appropriate" Pose

The significance of World War II to Japan's postwar technological dynamism – hence to its industrial accomplishment – seems to have been neither sufficiently stressed nor carefully studied. Yet the wartime cutoff from the mainstream of technological developments in the West had many important effects on Japan's subsequent effort – and efficiency – in catching up with the West, both technologically and industrially, in the postwar period.

In the first place, Japan's capacity to absorb and master industrial technology must have advanced enormously as a result of its own wartime research effort. Indeed, it was the first time that formal research programs were systematically organized and that researchers from different institutions collaborated in crash research activities – for the army, navy, and air force. Many technological advances, if in the nature of improvements, were made without the direct help of foreign experts, although they were mostly the results of attempting to copy the technological breakthroughs made in the West. For example, nylon and vinylon were successfully produced, though on a limited scale; so was penicillin – mainly with the help of patent documents and other published information. In fact, the phenomenal growth of Japan's synthetic fiber, pharmaceutical, and shipbuilding industries in the early postwar period owed a great deal to their wartime research experiences. Many successful postwar ventures in the production of radios, sewing machines, optical products, and other precision instruments were started by those who once were engaged in wartime research activities. Sony is the prime example in this respect.*

Secondly, during and especially in the years immediately after the war, Japan was compelled to take stock of its own technological capacity to find out in which industrial sectors and by how far it was behind. A variety of government surveys, including those conducted by the general headquarters of the Allied powers in connection with Japan's reparation programs, presented the overall picture of Japan's industrial and technological level as compared to that of the West.[7] These surveys were extremely useful exercises: any country that intends to learn from the outside world must clearly understand its own technological shortcomings in the first place. Thus comprehensive technological assessment, not in the currently used sense of analyzing the possible impact of new technology on society but in the sense of assessing a country's own technological capacity, is no doubt a prerequisite to the ability to absorb and master advanced technology. This type of technological stocktaking is still currently in practice. The Ministry of International Trade and Industry (MITI) has most recently released a survey indicating the sectors in which Japan still trails and those in which it leads the West.[8]

Thirdly, while Japan remained cut off from access to Western technology during and immediately after the war, many inventions

---

*Masaru Ibuka "had been chief engineer for Japan Precision Instrument Co., his own private measuring instruments company during the war, supplying the military with vacuum-tube voltmeters and a variety of other precision instruments. . . . During the war telephone conversations in China were being monitored, and Ibuka developed an audio frequency generator that, at 2,000 cycles, could be applied to a telephone system to ensure privacy of conversation. The military bought this and, further, interested in his theories, encouraged him to design a thermo-guidance system for bombs that would direct themselves to a source of heat; the system never reached completion, but during the project he met Akio Morita, a skinny twenty-three-year-old naval lieutenant, a trained physicist, who would later join his little firm in Tokyo." Nick Lyons, *The Sony Vision* (New York: Crown Publishers, 1976), pp. 3-4.

and innovations occurred in the West, increasing a stock of borrowable technology from which Japan would be able to choose once access was reopened. This was obviously an advantage for latecomer Japan. In introducing a given new product or process, for example, there were in many instances two or three alternative sources of technology for Japanese industry to choose from, so that it was able to compare them carefully and select the best possible one. The Japanese were also in a fortunate position for introducing the latest technologies in related industries simultaneously to bring about interindustry complementarities.[9] (Perhaps they stumbled upon this favorable development unintentionally when all sectors expanded at rapid rates, feeding on interindustry linkages.) Thus an expanded stock of borrowable technology gave them a wider selection and the simultaneous combinations of mutually reinforcing technologies.

In sum, although the implications of the wartime and early postwar period, during which Japan was left isolated from the mainstream of Western technological advance, are not fully explored and still need to be empirically studied, it was no doubt a uniquely important period in which Japan's technological absorptive capacity was substantially enhanced and in which Japanese industry was able to take stock of and consolidate its own technological capacity, as well as to wait for a stock of borrowable technologies to increase; it was the important preparatory stage of technological dynamism, the cumulative effects of which were later fully exploited in the three consecutive decades after 1950.

The 1950s: A Push for Heavy Industrialization

As soon as Japan regained political autonomy, it began to fill the technology gap by setting up a technology import program. The Foreign Exchange and Foreign Trade Control Law of 1949 and the Foreign Investment Law of 1950 formed the basis for administrative control of the acquisition of Western industrial arts. The Foreign Investment Committee was set up as the central body to screen applications for technology imports, which were classified into two classes, A and B, depending on length of contract duration. Class A covered those one year or longer and Class B those shorter than one year. The latter usually involved purchases of short-term or one-time technical assistance such as training, consultations, and acquisition of blueprints and other technical information.[10]

Since the acquisition of advanced technologies normally led to massive capital investments in new productive facilities, often opening up entirely new industries, the government control over technology imports was tantamount to the control over direction of industrialization and structural change. In fact, MITI initially did use technology import controls as a means of orchestrating the reconstruction and expansion of Japanese industry. Given its latecomer status, it was a relatively easy task for Japan to "pick winners." Priority was first given to the basic materials industries (such as metals – especially steel – and chemicals). Under the requirement of the Foreign Investment Law, a list of desired technologies to be imported was made public in 1951, and another in 1959, indicating priority sectors for modernization. (These lists, however, had little impact as guides; all the technologies considered desirable were aggressively acquired anyway by Japanese industry whether they were listed or not.) During the 1950s, for example, the government approved 1,029 class A technology import contracts, out of which 218 (21.2 percent) were secured by the nonelectric machinery industry, 217 (21.1 percent) by the electric machinery industry, 198 (19.2 percent) by the chemical industry, and 94 (9.1 percent) by the metal industry – these four industries as a whole accounting for 70.6 percent of the total. As might well be expected, the United States was the predominant supplier of these technologies, accounting for 65 percent; the second largest supplier was Switzerland, but it supplied only 8 percent. West Germany, Britain, and France were among other leading suppliers, each responsible for about 3 percent.[11]

It is worth stressing that Japanese industry secured key technologies mostly under licensing agreements. Direct foreign investments were limited in number and were set up, when they were approved by the government, as joint ventures – with only a handful of exceptions involving whole foreign ownership. Although modern machinery and equipment were imported, large-scale plant imports as we know them today were very rare, with the major exception of the power industry, which initially depended on the import of large thermal power stations each time generation capacity was upgraded. But even in this industry "MITI established a policy that only one new-type generator should be imported and the others be ordered from Japanese heavy electric machinery companies which would purchase the licenses for production. In this way, the three large makers of heavy electric machinery, Hitachi, Toshiba and Mitsubishi Electric, received a considerable boost."[12] In connection with this import substitution policy Hitachi and Toshiba each secured a license from General Electric (U.S.), Mitsubishi Electric from Westinghouse (U.S.), and the fourth largest, Fuji Electric, from Siemens (West German).[13] "No. 1 generator is foreign made, but no. 2 generator home-made" was the common pattern observed in Japan's power industry.

Similar import substitution was prevalent in other capital goods industries as well. Japanese industry produced as many capital goods as possible for itself under licensing agreements; imports of foreign-made equipment and machinery were kept minimal, usually as prototypes to be duplicated and improved upon at home. For this type of policy to be successful, Japan must obviously have had a highly developed absorptive capacity to master foreign technology in a short period. In order to do it on its own, indeed, Japanese industry had to devote a large amount of research money and effort to further develop, adapt, and commercialize imported technologies. Japanese borrowing was not a mere copying. Adaptive R & D was prerequisite to Japan's successful effort to indigenize foreign industrial arts. (About one-third of the research and development expenditures of Japanese industry over the period 1957-62 was, for example, used to "process" imported technologies.)[14]

The policy emphasis placed throughout the 1950s on the domestic production of basic materials and capital goods by acquiring foreign technology was intended, firstly, to modernize and expand the existing heavy industries, whose development Japan had initiated in the prewar years (for example, steel, machinery, chemicals, and shipbuilding), and secondly, to introduce brand new postwar industries (such as synthetic fibers, plastics, electronics, and petrochemicals). Most of these industries were capital intensive and resource-consuming and looked unpromising or even inappropriate for capital- and resource-indigent Japan – particularly in terms of the logic of the doctrine of comparative costs. Yet they were the very industries pushed for expansion by MITI and a group of high-growth advocates. In the words of former MITI Vice Minister Y. Ojimi,

> The Ministry of International Trade and Industry decided to establish in Japan industries which require intensive employment of capital and technology, industries that in consideration of comparative cost should be the most inappropriate for Japan, industries such as steel, oil refining, petro-chemicals, automobiles, aircraft, industrial machinery of all sorts, and electronics, including electronic computers. From a short-run, static viewpoint, encouragement of such industries would seem to conflict with economic rationalism. But from a long-range viewpoint, these are precisely the industries where income elasticity of demand is high, technological progress is rapid, and labour productivity rises fast.[15]

Some serious concerns, however, were expressed about such an audacious policy. H. Ichimada, then governor of the Bank of Japan, was, for example, highly critical of expanding Japan's steel production capacity (he said that huge steel mills would be covered with weeds) and manufacturing automobiles, which in his view Japan would be better off importing instead of producing. In fact, MITI itself was not so sure about the appropriateness of Kawasaki Steel's epoch-making decision to construct a huge integrated steel mill of a one-million-ton capacity in 1952.

The concern of the Bank of Japan and MITI about creating overcapacity, however, quickly became unwarranted because of unexpected favorable developments. Firstly, the Korean War and a subsequent export boom provided the first necessary stimulus to Japan's drive for modernization. About 70 percent of special procurement demands during the Korean War were centered on metals, machinery, and textiles.[16] Without such a dramatic surge in effective demand Japanese industry would not have invested so boldly in new technologies and productive facilities. Secondly, by concentrating on the introduction of modern technologies in the heavy and chemical sectors, the very technologies that would call for large-scale capital investments, Japanese industry unwittingly generated self-supportive demands for its increased output by way of the familiar income multiplier and accelerator mechanisms: massive capital investments expanded income, which in turn stimulated more investment, which again begat more income. In other words, "investment-creating investment" occurred with the continuous injection of new technologies as a vital stimulus. In this process, moreover, one industry after another experienced a "product-monopoly busting" phenomenon,* in which the first mover's monopolistic position, secured with a new foreign technology, was quickly challenged by competitors who introduced alternative technologies. Technological assimilation was thus closely linked with intercompany competitive dynamism in Japanese industry.

This competitive dynamism owed much to strong oligopolistic rivalries among Japan's leading industrial groups known as *keiretsu* (a much looser – and more pragmatic – affiliation of independent companies either vertically or horizontally interrelated to each other in a mutually supportive conglomeration than the old *zaibatsu*, which were feudalistically controlled by wealthy families). They strove to outperform each other in moving into new growth sectors. Intergroup oligopolistic competition never allowed Japan's growing markets, however protected from foreign competition, to be dominated by slumbering domestic monopolists. The existence of the industrial groups and the long-term capital finance made available by group-affiliated banks made Japanese companies quite willing to gamble on new growth sectors.

All in all, throughout the 1950s the hand of government was clearly seen promoting Japan's key industries, such as iron and steel, petrochemicals, shipbuilding, and heavy electric machinery, with the help of imported technologies. The First Rationalization Plan (1951-54) for the steel industry promoted blast furnace enlargements and sintering equipment improvements and the introduction of strip mills (continuous rolling); under the Second Rationalization Plan (1955-60), MITI required that every newly built mill have a capacity of one million tons and be as fully automated as possible.[17] During the second period the Linz Donawitz basic oxygen process was quickly introduced to replace the open hearth process.[18] For

---

*Takafusa Nakamura, for example, observes: "In order to expand production and increase sales, advances into new fields were planned unceasingly. As a result, 'product-monopoly busting' developed among steelmakers, who embarked on the production of special products monopolized by other companies. . . . In synthetic fibers, Toyo Rayon and Nippon Rayon maintained their dominant position in nylon while Teijin and other early producers held their lead in Tetron, but the late-coming companies also moved into nylon and took up the production of polyesters, competing with the original five companies in the import of technology." See T. Nakamura, *The Postwar Japanese Economy* (Tokyo: University of Tokyo Press, 1981), p. 65. A detailed study of technological assimilation in Japan's synthetic-fiber industry is presented in Ozawa, "Government Control."

petrochemicals MITI worked out detailed plans and set standards for the size of individual plants by adopting "the policy of refusing, in principle, authorization of any plant of a capacity of less than 300,000 tons."[19] The shipbuilding industry was fostered by the Ministry of Transportation's annual shipbuilding programs and a special loan facility for shipowners. The Western technologies of electric welding, block construction, and automatic gas cutters were adopted at Japanese shipyards.[20] The heavy electric machinery industry was, as explained earlier, supported by the government policy of encouraging the power industry to procure homemade generators.

In short, protection, technological absorption, the investment-induced growth of internal markets, and *keiretsu*-based competitive dynamism were the major features of Japan's industrial reconstruction and expansion throughout the 1950s, centered on heavy and chemical industries. The consumer durables sector, particularly electronics and automobiles, also started to import the latest technologies very actively in the late 1950s. But these industries' decisive expansion had to wait until the 1960s. Interestingly enough, they were the industries in whose growth the government played a relatively insignificant role.

The 1960s: The Advent of the Age of Mass Consumption

The 1960s saw an increased introduction of foreign technologies related to durable consumer goods and their mass production methods and an eagerness on the part of Japanese enterprises to learn Western, particularly U.S., techniques of management, a zeal popularly called "a management boom" in Japan. That decade also witnessed another feverish move of Japanese enterprises to set up R & D centers largely patterned after U.S. corporate research facilities.

With a phenomenal rise in personal income as the result of high economic growth, especially in the second half of the 1950s, Japanese consumer spending exploded. From 1955 to 1959, for example, the average personal income for both urban workers and farmers nearly doubled in real terms.[21] The age of mass consumption, epitomized by such popular slogans as "the consumer is the king" and "consumption is a virtue," arrived in Japan, starting in the early 1960s. For example, TV sets, which were used in only 16 percent of urban households in 1958, were used in as many as 91 percent by 1963.[22] The popular consumer durables in the early 1950s were relatively low-income products such as radios, fluorescent lamps, bicycles, electric fans, and sewing machines. The latter 1950s saw a growth in the demand for monochrome TV sets, washing machines, and automatic rice cookers. The demand for more sophisticated high-income goods such as color TVs, refrigerators, air conditioners, and automobiles expanded phenomenally throughout the 1960s, but especially in the latter years of the decade.*

Consumer demand for clothing similarly underwent a qualitative change: synthetic fiber or blended fiber products became increasingly popular as many of them had a permanent press feature. And tastes for fashion-oriented clothing, too, developed significantly. Many of these consumer goods are time-saving in nature for housewives and created more leisure time. On the average, the Japanese housewife spent 11 hours per day on houshold chores before the war but less than 8 hours at the end of the 1960s.[23] Innovations in the consumer goods sector, therefore, in turn created new demands for leisure-related goods and services. These revolutionary changes in the pattern of consumption were also reflected in a sharp rise in the number of foreign technology imports that were specifically related to the consumer goods and leisure-related industries.†

At some risk of oversimplification, the technological drive of the 1950s can be characterized basically as "supply-push" (since the major stimulus to technological progress came from the ready availability of advanced technologies already market-tested in the West, particularly in the capital goods sector, and from the policy emphasis of the government on building key industries); that of the 1960s, by contrast, may be distinguished as "demand-pull" (for it was the domestic consumer demand that played a key role in encouraging manufacturers to introduce a succession of new products,

---

*An explosive rise in the demand for automobiles continued into the 1970s. In 1965 only 5.7 percent of Japanese households owned cars, but the ratio rose to 22.0 percent in 1970 and 40 percent in 1975. See Taku Oshima, *Jidosha Sangyo* (The Automobile Industry) (Tokyo: Toyo Keizai, 1980), p. 139.

†Class A technology import contracts for packaging and distribution, furniture, cosmetics and sundries, dress designs, and leisure activities increased from 58 in 1963 (accounting for 10 percent of the total technology imports) to 216 in 1971 (accounting for 14 percent). Ozawa, *Japan's Technological Challenge*, p. 122.

many of which had some important innovative characteristics specifically designed to satisfy the Japanese consumer markets). The prime example is subcompact cars, which are economical in gas consumption and suitable for the narrow streets in Japan, and small TV sets, which are space-saving.

Many Japanese innovations (which are based on foreign inventions but perfected with modifications and improvements) such as small cars, transistorized radios, miniaturized TV sets, pocket-size calculators, cassette recorders, fully automatic cameras, and, most recently, home video tape recorders, have all been supported by strong domestic demand, which provided the basis for scale economies and export competitiveness. It was also the fastidious Japanese consumer who demanded high quality, great variety, many convenience features, and after-sale service, which all came to be translated into and embodied in the consumer goods by fiercely competing Japanese manufacturers. The nature of Japan's technological dynamism, notably in consumer electronics, optics, and automobiles, can never be adequately explained unless one takes into consideration the vital role played by the huge internal market, the demand of Japanese consumers for quality and service, and the vigor with which the manufacturers vie with each other in catering to the whims of their customers.

The emergence of affluent mass consumption markets at home and Japan's effort to export consumer durables to high-income Western countries, notably the United States, in turn necessitated market-oriented R & D. It was against this backdrop that "posh" research centers were set up one after another by leading Japanese manufacturers in the first half of the 1960s – again in competition with one another by emulating U.S. corporations. This movement occurred pari passu with a rapid rise in R & D expenditures, especially in the latter half of the 1960s, when Japanese enterprises seriously stepped up research efforts and "indigenous technological development (Jishu Gijutsu Kaihatsu)," became a catchphrase. R & D expenditures jumped from about ¥250 billion in 1965 to more than ¥800 billion in 1970, more than tripling over the five-year period.[24] The ratio of R & D funds to sales at the corporate level also increased from 1.0 percent to 1.4 percent, on the average. Yet Japan's R & D activities were still mostly directed toward commercializing imported technologies. In fact, as soon as the technology import controls were substantially liberalized in June 1968, the number of technology purchase contracts (class A) soared from 638 in 1967 to 1,061 in 1968 and continued to climb to a record high of 1,916 in 1972, a strong indication of Japan's continued dependence on foreign technologies (see Appendix Figure 1).

Japan's borrowing of industrial knowledge from the West was not confined to the realm of production alone: Japanese business managers were also eager learners of Western organization and management techniques, if not philosophies. In fact, such now familiar terms as automation, marketing, human relations, and productivity (*seisansei*) are all new jargon introduced into the Japanese lexicon only in the postwar period. In this connection, the Japan Productivity Center, a governmental organization established in 1959 with the help of the United States (which contributed ¥2.3 billion to the total fund of ¥10 billion) played a pivotal role in providing Japanese businessmen with the opportunities to visit U.S. and European factories, business offices, and research centers in order to learn first-hand the secrets of Western management. Numerous seminars, lectures, and special guidance programs were given to visiting Japanese managers, union leaders, and public officials.

It was also during the 1960s that the now famed practice of quality control (QC) circles took root in Japan. The basics of control charts and sampling inspection with the use of statistical principles was introduced into Japan by W. E. Deming, of the United States, in 1950, and the first seminar on QC was given by J. M. Juran, also of the United States, in 1954. Although the techniques and concept of statistical quality control and total quality control, originally developed in academia in the United States, were thus imported, the practical QC techniques *at the workshop level* were developed to take advantage of Japan's special workplace environment by Japanese corporations themselves.

Throughout the 1960s Japan's efforts to restructure its industry toward heavy and chemical industries also continued incessantly, now including a phenomenal expansion of consumer-oriented, high value-added manufacturing sectors. The ratio of output value of heavy and chemical industries to total value of manufactures rose from 42.2 percent in 1955 to 60.6 percent in 1965 and to 68.6 percent in 1970.[25] But the very success of this effort, combined with drastic changes in the external industrial environment, would necessitate a reorientation of Japan's technological and industrial drive in the 1970s.

The 1970s: Efforts to Cope with Social Needs and External Constraints

By the end of the 1960s, the industrial environment for Japan had undergone considerable change. Numerous constraints emerged, decreasing the free rein given to industrial expansion under the probusiness, growth-oriented policies of the government. One major weakness of the industrial and technology policies pursued during the 1950s and 1960s from the social point of view, if not from that of private industry, was a total disregard of the externalities or social costs of industrialization and modern technology, especially in the form of pollution and urban congestion. These externalities were further worsened by the very emphasis Japan gave to the development of the heavy and chemical industries – highly resource-consuming and pollution-prone as they were – and by the very success in achieving this goal. This was one of the most serious internal problems Japan had to solve in the early 1970s.

Ironically, moreover, this drive toward heavy industrialization made Japan one of the world's most resource-poor economies, relatively a heavier consumer of mineral and energy resources than either the United States or West Germany.[26] And this development led naturally to a point where Japan was vulnerable through heavy dependence on overseas resources. This was emphasized by the 1973 Arab oil embargo, for at that time, as much as 76 percent of Japan's primary energy requirements were met by oil, and almost the entire amount came from abroad. It therefore became imperative for Japan to develop technologies that would reduce the consumption of oil and increase the supply of nonoil energy resources.

Another technological imperative that emerged toward the end of the 1970s was of innovating labor-saving methods of production. Japan began to find itself short of labor, particularly young factory workers, in the mid-1960s. Rapid economic growth had by then transformed the Japanese economy into a labor-scarce and capital-abundant one. As a result, wage rates soared and labor shortages became a serious bottleneck throughout the economy, but especially in labor-intensive, traditional light manufacturing sectors, where small- and medium-sized firms are concentrated. A sudden sharp appreciation of the yen in the early 1970s also weakened the export competitiveness of Japan's low value-added light manufactures.

With these drastic internal and external changes as a background, the government announced an epoch-making economic policy, a proposal made by MITI's Industrial Structure Council, at the start of the 1970s. The new policy emphasized a reorientation of industrial structure away from pollution-prone and natural-resource-consuming heavy and chemical industries and toward clean and knowledge-intensive industries, whose products could compete in the world market in terms of quality, variety, and sophisticated design rather than primarily in price.

The emergence of these technological imperatives was clearly reflected in the changing priorities of research activities in the private sector, which were revealed in a survey made by Japan's Science and Technology Agency in 1979 on goals of R & D (Figure 7.1). The achievement of internal needs (that is, improvement of product quality and production efficiency, enhancement of product convenience and comfort, and perfection of mass production techniques) was the dominant theme of R & D throughout the 1950s and 1960s (in both decades, it accounted for more than two-thirds of research goals). Yet with the arrival of the 1970s, research efforts to respond to external constraints (such as labor shortages, high resource prices, and environmental protection) gained in importance, explaining close to one-half of research objectives. And these new efforts are expected to intensify further throughout the 1980s.

The trend of technology imports in the 1970s similarly mirrored Japan's new technological imperatives. Technology import contracts related to prevention of environmental deterioration (that is, pollution control, waste water treatment, and noise abatement) all of a sudden increased in the early 1970s, taking more than 10 percent of the total number of technology imports.[27] Toward the end of the decade, however, this type of technology acquisition steadily declined as advanced technologies, both imported and indigenously developed, were successfully put into use with the result of considerable improvement in environmental conditions. Japan's automobile industry, in particular, was successful in quickly introducing low-emission engines, as seen in the Honda Motor Company's development of controlled vortex combustion chamber (CVCC) engines.

Japan's imports of foreign technologies related to the development of energy (especially nuclear power generation and the extraction and refining of oil and natural gas) began to rise sharply after

FIGURE 7.1
Goals of R & D in Japanese Industry

Internal motivation (responding to internal needs)
A. Improvement of quality and efficiency
B. Enhancement of convenience and comfort
C. Mass production
External motivation (responding to external constraints)
D. Environmental protection
E. Safety
F. Labor saving
G. Physical resource saving
H. Energy saving

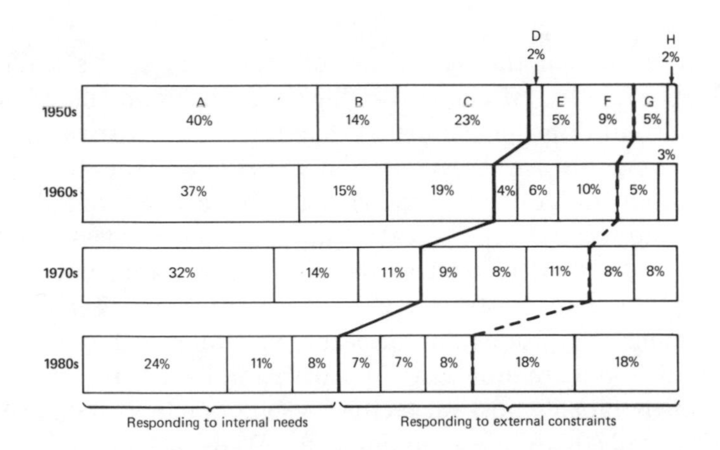

Source: Adapted from Japan, Science and Technology Agency, *Kagaku Gijutsu Hakusho* [White Paper on Science and Technology], 1980, p. 113. The original data came from a questionnaire survey conducted by the agency in 1979.

the oil crisis of 1973. These technologies were more often imported under class B than class A contracts. In 1978, the year in which the largest number of energy-related technology imports was recorded, they accounted for 12.2 percent of all class B and 3.6 percent of all class A contracts.[28]

As far as labor-saving devices are concerned, Japanese companies depended on their own technologies rather than on imported ones, although Japan did initially purchase "seed technologies" from the West. For example, in 1968 Kawasaki Heavy Industries was the first

Japanese enterprise to produce industrial robots under the licensing agreement with Unimation of the United States, and in 1969 Kobe Steel followed suit by acquiring a robot technology from a Norwegian company.[29] According to an estimate made by the American Robot Association, in 1980 Japan had 14,246 variable sequence or more sophisticated robots in operation, the United States 4,100, West Germany 1,130, and the United Kingdom 371.[30]

Technology Imports and Adaptations during the Three Decades: An Overall Assessment

I have examined the nature and direction of Japan's technological efforts and the influences on them of different sets of macroeconomic conditions prevailing during the three postwar decades, with the long historical assimilation experience (the tradition of learning from the outside world), and the century-old drive for industrial modernization (to catch up with and surpass the West) as underlying forces, as well as the wartime research autarky as a valuable prelude to postwar technological progress. Technological adoptions and adaptations do not occur in a situational vacuum: they are induced – and are molded – by existing industrial milieus.

Although Japan's postwar technological miracle would have been impossible had there been no imports of advanced industrial arts from the West, Japanese industry itself exerted a great deal of effort to adapt and assimilate imported technologies. This assimilative effort initially stimulated adaptive R & D, which later turned more original in orientation. There has been a constant rise in R & D expenditures: as a ratio of national income they were a mere 0.8 percent in 1953 but rose to 1.42 percent in 1960, 2.0 percent in 1970, and 2.4 percent in 1980. (The same ratio for the United States was 2.6 percent in 1980 and for West Germany 2.7 percent in 1979).[31] Technology imports, which reached their peak in 1973 (2,450 contracts, both class A and class B) have shown a declining trend ever since. Similarly, the ratio of royalty payments to R & D expenditures, which had kept rising until 1960, reaching its highest level of 18.5 percent, has thereafter been decreasing: it came down to 5.1 percent in 1980 (see Figure 7.4). In terms of the number of patent applications, Japan was well behind the Soviet Union, the United States, West Germany, and the United Kingdom in 1960, but it has ranked first ever since 1973.[32] All these developments clearly testify to a rapid rise in the level of Japan's industrial technology. No doubt, technology imports have been the most significant stimulant to the development of Japan's own R & D industry.

TECHNOLOGY OUTFLOW FROM JAPAN

What forms do Japan's exports of technology take? What are the macroeconomic forces behind them? These are the two main questions to be considered in this section.

In a nutshell, Japan's exports of technology are at the moment greater to developing countries than to other industrialized countries. The former mostly take the form of direct foreign investment and plant exports (particularly on a comprehensive turnkey basis), whereas the latter are made primarily under licensing agreements, although direct foreign investment is on the rise.

One basic factor underlying technology outflows from Japan, whether to developing or industrialized countries, is the phenomenal structural changes in the postwar Japanese economy. In fact, technology outflows have been a concomitant phenomenon of these changes.

Industrial Restructuring and Technological Spinoffs

As seen earlier, with an incessant infusion of advanced technologies Japanese industry has gone through a swift industrial restructuring, initially, away from low value-added light manufacturing and toward higher value-added heavy and chemical industries – and more recently, away from the resource-consuming heavy and chemical sectors and toward knowledge-intensive, pollution-free sectors.

When we talk about structural changes it is useful to distinguish, as the Japanese government does, between basic materials industries (that is, metals, metal products, basic chemicals, ceramics, and textiles), which are both relatively more raw material intensive and energy-consuming, and fabricating-assembly industries (electric and nonelectric machinery, transport equipment, and precision machinery), which are relatively more knowledge intensive and higher in value added. The former type of industry produces mostly homogeneous and mature products, whereas the latter produces, on the whole, highly differentiated products for which nonprice factors such as advertising and customer services are important.

The immediate objective of Japanese industry during the 1950s was to reconstruct and expand both basic materials and fabricating-assembly industries. Yet as industrialization proceeded, it was the fabricating-assembly manufacturing activities (the demand for whose output is relatively more income elastic) that came to be stressed in the 1960s and later. In fact, the fabricating-assembly sector, which had a mere 17 percent share of manufactures in 1955, came to account for as much as 62 percent in 1980, while the basic materials sector, which had about a 50 percent share in 1955, had by 1980 shrunk to 21 percent.[33]

Japan's swift industrialization (structural reorientation toward higher value-added activities) entailed a sharp rise in wages. In the latter half of the 1960s there appeared an acute shortage of young workers, particularly high school graduates, willing to be employed as factory workers. There was in fact an absolute decline in the number of 15-to-19-year-olds in the labor force at the very time when the demand for workers in this age-group was on the rise, since they were highly adaptable, both physically and mentally, to modern factory production, especially in fabricating-assembly manufacturing. As a way to escape from labor shortages at home, Japanese producers began to look for manufacturing locations in labor-abundant countries; also, it so happened that many developing countries in Asia in particular were then eagerly inviting foreign capital investment in exactly those sectors where Japan began to lose comparative advantage. It was against this background that Japan's initial technology transfers in the postwar period took place, mostly in the form of direct manufacturing investments in labor-abundant developing countries.

As detailed elsewhere, Japanese manufacturing ventures exhibit characteristics distinct from those of their Western counterparts. First, more than 70 percent of their manufacturing investments are located in developing countries and are centered on low-technology, standardized products such as textiles and metal products, on the low-technology, standardized end of electric and electronics appliances (for example, batteries, fans, radios, and TV sets) and basic chemicals (for example, paints, plastics, and agricultural chemicals).

Second, they are actively participated in by small- and medium-size firms, and many of them are therefore small in scale of operation. This characteristic is most conspicuous in Asia: about half of Japanese manufacturing ventures there are set up by this class of firms. Third, Japanese ventures have a relatively high propensity to form joint ventures with local interests by accepting minority ownership, especially in developing countries.[34]

Thus the manufacturing technologies transferred to developing countries through Japanese ventures are by and large labor-intensive intermediate technologies (perhaps obsolete by Japanese standards in some instances but commercially still appropriate for the factor endowment conditions of developing host countries) and are transplanted to the local community mostly in relatively small-scale joint ventures, in which Japanese partners are frequently minority owners. (There is no guarantee that technologies are actually transplanted into the local industrial community. But joint ventures are at least likely to create many more opportunities for local interests' involvement in management and production than wholly-owned foreign subsidiaries.)

Most recently, Japanese industry (particularly fabricating-assembly sectors) is actively setting up ventures to manufacture technology-intensive products (such as color TVs, integrated circuits, and automobiles) in other industrialized countries, particularly the United States and European countries. These are mostly designed to avoid trade frictions (caused largely by the very success of the Japanese economy in developing the high-technology end of its industrial structure as new export industries). Yet in contrast to those in developing countries, these ventures are normally wholly owned by Japanese corporations and engaged, at the moment, in assembly type operations with key parts and components mostly imported from Japan. Technological transfers to the developed host countries through the medium of direct investment, if any, are therefore only indirect and perhaps insignificant at the moment. Licensing agreements are a more direct conduit of Japan's technology transfers to advanced countries.

Plant Exports as a Technology Transfer Agent

In addition to many successful exports such as consumer electronics products and automobiles, Japan has recently attained another phenomenal expansion in its exports of plants that can produce electric power, steel, aluminum, fertilizers, petrochemicals, and textiles. In 1970 Japan ranked fifth as a plant exporter among the Organization for Economic Cooperation and Development countries after West Germany, the United States, the United Kingdom, and France but soon overtook the United Kingdom and France during the second half of the 1970s. Plant exports now account for about 10 percent of the total value of Japanese exports, having emerged as one of the country's major exports next to automobiles and electronics products. In 1980, 67.5 percent of Japan's plant exports were sold to developing countries and 21.1 percent to communist countries.[35]

This success is due to a host of factors. Perhaps the most significant is, again, the swift industrial restructuring Japan has experienced during the postwar period. The industrial machinery sector, in particular, has accumulated a great deal of skill and experience in improving plant layouts, material handling techniques, machinery, and equipment. Many quality control devices are built into plants. All the modernization experience and skill Japanese industry has gained are embodied in the capital goods it now produces and exports. This feature is said to be a great attraction to Third World countries, which are eager to learn from Japan's recent experience of industrialization.

Plant exports accompany transfers of both product and process technologies, both "hardware" and "software," simultaneously, particularly when comprehensive turnkey contracts that include training of operatives for production and maintenance are involved. Infrastructural building that requires the training and hiring of local construction workers is also a usual feature of comprehensive contracts. The "software" component often explicitly takes the separate form of technology assistance contracts for which royalties and fees are received. Receipts from these contracts are included in Japan's official statistics on technology exports.

Figure 7.2 shows Japan's receipts from technology assistance contracts by region and industry in 1980. Interestingly enough, relatively large proportions of receipts from developing regions are secured in the construction industry (as much as 74.4 percent from the Middle East, 9.5 percent from Asia, and 8.6 percent from Latin America), whereas this industry is insignificant so far as receipts from advanced regions are concerned. This characteristic is closely

FIGURE 7.2
Japan's Receipts from Technology Exports, 1980
(in million yen)

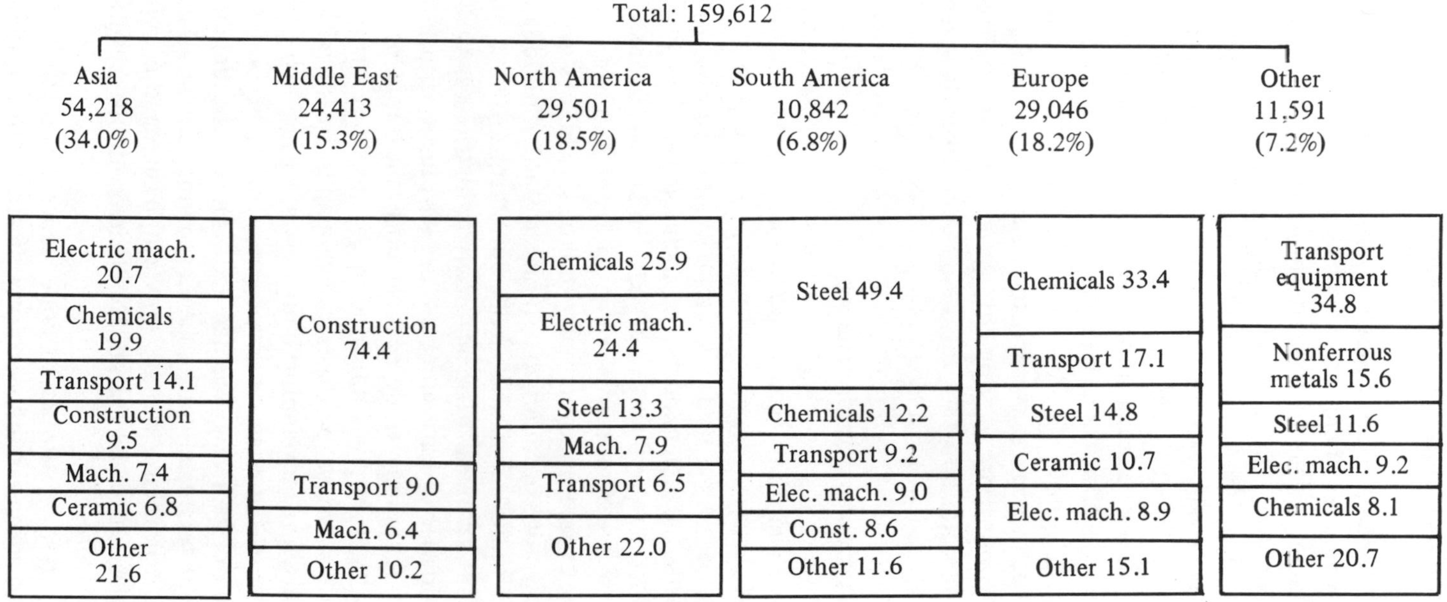

Total: 159,612

Asia 54,218 (34.0%)	Middle East 24,413 (15.3%)	North America 29,501 (18.5%)	South America 10,842 (6.8%)	Europe 29,046 (18.2%)	Other 11,591 (7.2%)
Electric mach. 20.7	Construction 74.4	Chemicals 25.9	Steel 49.4	Chemicals 33.4	Transport equipment 34.8
Chemicals 19.9	Transport 9.0	Electric mach. 24.4	Chemicals 12.2	Transport 17.1	Nonferrous metals 15.6
Transport 14.1	Mach. 6.4	Steel 13.3	Transport 9.2	Steel 14.8	Steel 11.6
Construction 9.5	Other 10.2	Mach. 7.9	Elec. mach. 9.0	Ceramic 10.7	Elec. mach. 9.2
Mach. 7.4		Transport 6.5	Const. 8.6	Elec. mach. 8.9	Chemicals 8.1
Ceramic 6.8		Other 22.0	Other 11.6	Other 15.1	Other 20.7
Other 21.6					

Source: Japan, Science and Technology Agency, *1982 Kagaku Gijutsu Hakusho* [White Paper on Science and Technology], Tokyo, 1982, p. 237.

related to Japan's plant exports that accompany construction contracts. Heavy and chemical industries – chemicals, iron and steel, electric machinery, and transport equipment – are all active exporters of technology. Latin America is especially an eager absorber of Japan's steel technologies in connection with plant imports. North America, especially the United States, purchases technologies under license for chemicals, electric machinery, and steel (direct investments are also becoming important conduits for electric machinery in particular as Japanese companies set up factories in the U.S. electronics industry). Somewhat similar patterns are discernible for Europe. Technology sales to Asian countries by electric machinery, chemicals, transport equipment, and other industries are made mostly in conjunction with Japanese manufacturing investments.

Sectoral Technology Trade Balance

Perhaps a more interesting picture emerges when we look at the balance of Japan's technology trade in terms of both receipts and payments, as shown in Table 7.1. The construction industry is the most substantial net exporter of technology: its receipts are close to ten times its payments. The basic materials industries have also emerged or are about to emerge as a net exporter of technology (with the exception of metal products), whereas the fabricating-assembly sector continues to exhibit substantial deficits, although its receipts are quite substantial in absolute terms.

The fact that the basic materials industries (as well as the construction industry) have emerged as a surplus – or a near surplus – sector in technology trade reflects both the rapid past growth and the maturity of these industries in Japan. They are now the very industries for whose products the demands are sluggish worldwide, but which many developing countries are eager to establish as the key industries for their development efforts (the same effort Japan itself made only a few decades ago). Hence the demand for plant exports and technical assistance is growing. Japan's basic materials industries, for their part, are now emphasizing the export of their technologies, since the market for their products, both at home and abroad, is depressed.

In contrast, a substantial dependence on imported technologies in the fabricating-assembly sector indicates both the vigorous growth

TABLE 7.1
Japan's Balance of Technology Trade by Sector, 1980
(in million yen)

	A. Receipts	*B. Payments*	*A/B*
Total	159,612 (100.0%)	239,529 (100.0%)	0.66
Manufacturing	133,274 (83.5)	233,185 (97.4)	0.57
Basic materials			
Textiles	3,169 (2.0)	2,233 (0.9)	1.42
Chemicals	31,876 (20.0)	39,252 (16.4)	0.81
Ceramics	7,989 (5.0)	9,612 (4.0)	0.83
Iron & steel	17,856 (11.2)	8,023 (3.3)	2.23
Nonferrous metals	3,663 (2.3)	3,690 (1.5)	0.99
Metal products	1,221 (0.8)	4,440 (1.9)	0.28
Fabricating-assembly			
Nonelectric machinery	9,621 (6.0)	30,209 (12.9)	0.32
Electric machinery	23,045 (14.4)	61,676 (25.7)	0.37
Transport equipment	21,758 (13.6)	40,274 (16.8)	0.54
Precision machinery	873 (0.5)	2,948 (1.2)	0.29
Other manufactures	12,203 (7.6)	30,828 (12.9)	0.39
Construction	25,399 (15.9)	2,707 (1.1)	9.38
Other services	938 (0.6)	3,637 (1.5)	0.26

Source: Adapted from Japan, Science and Technology Agency, *1982 Indicators of Science and Technology*, Tokyo, 1982, pp. 118-21.

stage of these industries in Japan and the capacity of Japanese firms to capitalize on and successfully commercialize the latest technologies available in the West. They are in fact currently highly R & D-intensive industries, and Japanese firms themselves are investing heavily in research activities. In 1980 R & D expenditures as a percentage of net sales stood at 1.73 for the manufacturing sector as a whole, but every single fabricating-assembly industry's ratio exceeded the overall average (1.9 percent for nonelectric machinery, 3.7 percent for electric machinery, 2.3 percent for transport equipment, and 3.0 percent for precision machinery),[36] an indication of the relatively youthful stage of these industries.

CONCLUSION

My purpose in this chapter has been to discuss technology inflows to and outflows from postwar Japan. In so doing I have identified two sets of macroeconomic determinants of Japan's technological dynamism. One consists of Japan's long historical experience and faculty to assimilate alien cultures, its continuity as a national effort to catch up with the West ever since the mid-nineteenth century, and its brief technological autarky during World War II. Comingled with these historical forces, sufficiently distinct macroeconomic determinants appeared in each of the three postwar decades considered in this chapter, though some factors were common to the entire period. In particular, a continuous drive for structural adaptations to changes in technology and demand was the underlying theme of Japanese industry throughout the postwar period. In essence, technology inflows to and outflows from Japan have been the means, as well as the results, of facilitating the structural upgrading of the Japanese economy.

FIGURE 7.3
Japan's Imports of Technology, 1950-81
(in number of purchase contracts)

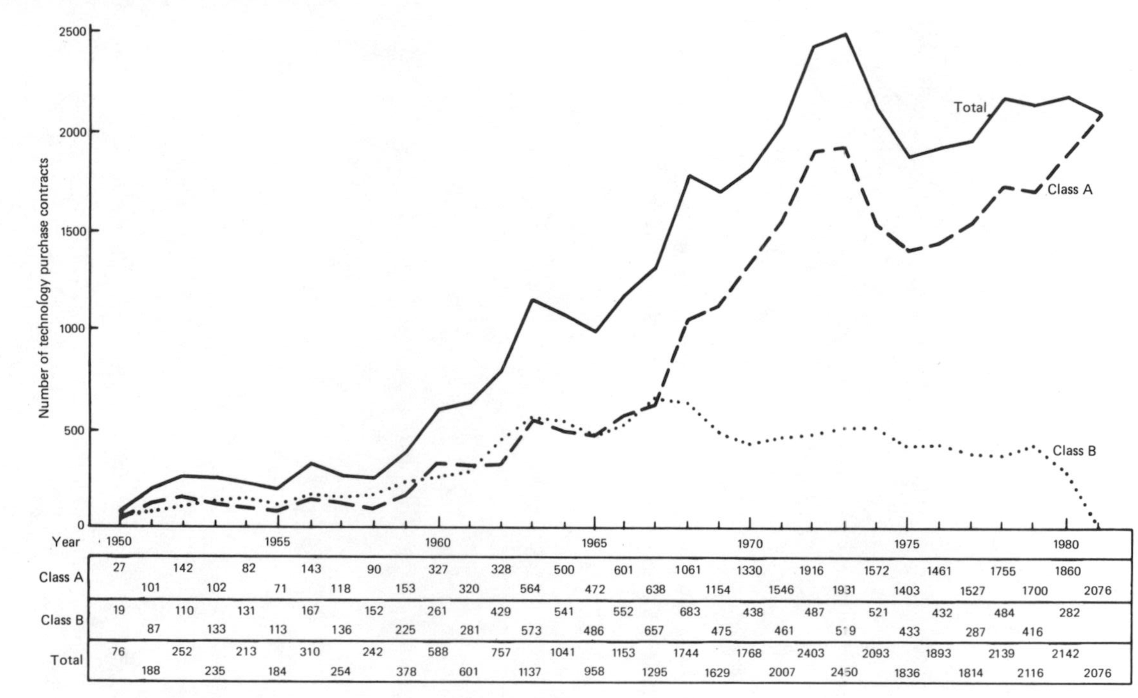

Year	1950	1951	1952	1953	1954	1955	1956	1957	1958	1959	1960	1961	1962	1963	1964	1965
Class A	27	101	142	102	82	71	143	118	90	153	327	320	328	564	500	472
Class B	19	87	110	133	131	113	167	136	152	225	261	281	429	573	541	486
Total	76	188	252	235	213	184	310	254	242	378	588	601	757	1137	1041	958

Year	1966	1967	1968	1969	1970	1971	1972	1973	1974	1975	1976	1977	1978	1979	1980	1981
Class A	601	638	1061	1154	1330	1546	1916	1931	1572	1403	1461	1527	1755	1700	1860	2076
Class B	552	657	683	475	438	461	487	519	521	433	432	287	484	416	282	
Total	1153	1295	1744	1629	1768	2007	2403	2450	2093	1836	1893	1814	2139	2116	2142	2076

Note: The A and B classifications were abolished at the end of 1980.

Source: Adapted from Japan, Science and Technology Agency, *1981 Gaikoku Gijutsu Donyu Nenji Hokoku* (1981 Annual Report on Foreign Technology Acquisition), Tokyo, 1982, p. 19.

FIGURE 7.4
Ratio of Technology Import Payments to R & D Expenditures, 1953-80

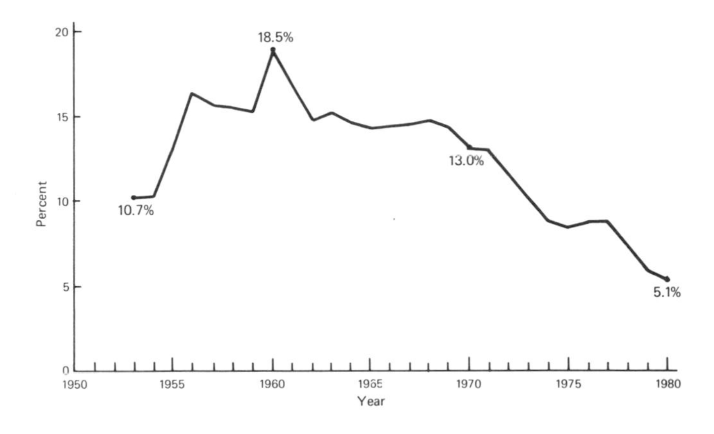

Source: Compiled from Japan, Science and Technology Agency, *1980 Kagaku Gijutsu Hakusho* (White Paper on Science and Technology), p. 110.

NOTES

1. Japan, Science and Technology Agency, *1980 Kagaku Gijutsu Hakusho* (White Paper on Science and Technology), Tokyo, 1980, p. 17.

2. The importance of historical momentum in Japan's technological assimilation is stressed in Terutomo Ozawa, "Technology Transfer and Japanese Economic Growth in the Postwar Period," in Robert Hawkins and A. J. Prasad, eds., *Technology Transfer and Economic Development* (Greenwich, Conn.: JAI Press, 1981), pp. 91-116.

3. Royal Academy of Arts, *The Great Japan Exhibition. Art of the Edo Period 1600-1868*, London 1981-82, p. 294. One can witness, for example at the British Museum, some convincing evidence of their ability to assimilate foreign industrial arts. Several ingenious modifications were made on the clock face to adapt to the peculiar time system used in Edo period Japan.

4. For emphasis on the Japanese ethos as the major factor of Japan's economic success, see Michio Morishima, *Why Has Japan "Succeeded"? Western Technology and the Japanese Ethos* (Cambridge: Cambridge University Press, 1982).

5. A. Hall, "Epilogue: The Rise of the West," in C. Singer, ed., *A History of Technology* vol. 3. (Oxford: Oxford University Press, 1957), pp. 716-17, as quoted in Nathan Rosenberg, *Inside the Black Box: Technology and Economics* (Cambridge: Cambridge University Press, 1982), p. 12.

6. Solomon B. Levine and Hisashi Kawada, *Human Resources in Japanese Industrial Development* (Princeton, N.J.: Princeton University Press, 1980), p. 41.

7. The general headquarters, for instance, arranged a special survey on Japan's rayon industry. A report entitled "The Rayon Industry of Japan" was made public in 1946. See Terutomo Ozawa, "Government Control over Technology Acquisition and Firms' Entry into New Sectors: The Experience of Japan's Synthetic-Fibre Industry," *Cambridge Journal of Economics*, 4 (June 1980), pp. 133-46.

8. Japan, Ministry of International Trade and Industry, *Wagakuni Sangyo no Gijutsu Suijun narabini Gijutsu Kaihatsuryoku Suijun ni kansuru Ankeito Chosa* (A Questionnaire Survey on Japan's Technological Level and Technological Development Capacity), Tokyo, 1982.

9. The significance of complementarities as a source of technological synamism is stressed by Nathan Rosenberg. See Rosenberg, *Inside the Black Box*, pp. 56-62.

10. For an overview of Japan's technology import program, see Dan F. Henderson, *Foreign Enterprise in Japan: Laws and Policies* (Chapel Hill: University of North Carolina Press, 1973); Terutomo Ozawa, *Japan's Technological Challenge to the West, 1950-1974: Motivation and Accomplishment* (Cambridge, Mass.: MIT Press, 1974); and Merton J. Peck, with the collaboration of Shuji Tamura, "Technology," in Hugh Patrick and Henry Rosovsky, eds., *Asia's New Giant: How the Japanese Economy Works* (Washington, D.C.: Brookings Institution, 1976).

11. Japan, Science and Technology Agency, *1980 Kagaku Gijutsu Hakusho*, p. 22.

12. Johannes Hirschmeier and Tsunehiko Yui, *The Development of Japanese Business 1600-1980* (London: George Allen & Unwin, 1981), p. 310.

13. Hiromi Arisawa et al., *Nihon Sangyo Hyakunenshi* (100-year History of Japanese Industry), second vol. (Tokyo: Nihon Keizai Shimbun, 1977), p. 59.

14. Japan, Ministry of International Trade and Industry, *Gijutsu Doko Chosa Hokokusho* (Report on the Trend of Technology).

15. Y. Ojimi, "Basic Philosophy and Objectives of Japanese Industrial Policy," in Organization for Economic Cooperation and Development, *The Industrial Policy of Japan* (Paris: OECD, 1972), p. 15.

16. Arisawa et al., *Nihon Sangyo Hyakunenshi*, p. 27.

17. Nakamura, *The Postwar Japanese Economy*, p. 70.

18. For an analysis of Japan's ability to adopt oxygen steelmaking processes, see Leonard H. Lynn, *How Japan Innovates. A Comparison with the U.S. in the Case of Oxygen Steelmaking* (Boulder, Colo.: Westview Press, 1982).

19. Ojimi, "Basic Philosophy," p. 17.

20. Tuvia Blumenthal, "The Japanese Shipbuilding Industry," in Hugh Patrick, ed., *Japanese Industrialization and Its Social Consequences* (Berkeley: University of California Press, 1976), pp. 155-56.

21. Arisawa et al., *Nihon Sangyo Hyakunenshi*, p. 137.

22. Ibid., p. 150.

23. Japan, Science and Technology Agency, *1976 Kagaku Gijutsu Hakusho*, p. 185.

24. Ibid., annex p. 36.

25. Japan, Ministry of International Trade and Industry, Industrial Structure Council, *Hachijunendai no Sangyokozo no Tembo to Kadai* (The Prospects for the Industrial Structure of the 1980s), Tokyo, 1981, p. 139.

26. This development is discussed in Terutomo Ozawa, *Multinationalism, Japanese Style: The Political Economy of Outward Dependency* (Princeton, N.J.: Princeton University Press, 1979), pp. 169-75.

27. Computed from Japan, Science and Technology Agency, *1976 Indicators of Science and Technology*, Tokyo, 1976.

28. Ibid., p. 224.

29. Nikkan Kogyo Shimbun, *Roboto Sangyo Chizu* (Robot Industrial Map), (Tokyo: Nikkan Kogyo Shimbun, 1982), pp. 11-12.

30. Japan, Ministry of Labor, *1982 Rodo Hakusho* (White Paper on Labor), Tokyo, 1982, p. 65.

31. Japan, Science and Technology Agency, *1982 Indicators of Science and Technology*, Tokyo, 1982, p. 10.

32. Japan, Science and Technology Agency, *1980 Kagaku Gijutsu Hakusho*, p. 110.

33. Japan, Science and Technology Agency, *1982 Kagaku Gijutsu Hakusho*, p. 4.

34. For details of these characteristics, see, for example, Kiyoshi Kojima, *Direct Foreign Investment. A Japanese Model of Multinational Business Operations* (London: Croom Helm, 1978), and Ozawa, *Multinationalism*, chaps. 1-3.

35. Japan, Ministry of International Trade and Industry, *1982 Tsusho Hakusho* (White Paper on International Trade), Tokyo, 1982, p. 222.

36. Japan, Science and Technology Agency, *1982 Indicators*, p. 51.

REFERENCES

H. Arisawa, K. Yamaguchi, K. Hattori, T. Nakamura, T. Miyashita, and M. Sakisaka. *Nihon Sangyo Hyakunenshi* (100-year History of Japanese Industry), second vol. Tokyo: Nihon Keizai Shimbun, 1977.

T. Blumenthal. "The Japanese Shipbuilding Industry." In *Japanese Industrialization and Its Social Consequences*, edited by H. Patrick. Berkeley: University of California Press, 1976.

A. Hall. "Epilogue: The Rise of the West." In *A History of Technology*, edited by C. Singer, vol. 3. Oxford: Oxford University Press, 1957.

D. F. Henderson. *Foreign Enterprise in Japan: Laws and Policies*. Chapel Hill: University of North Carolina Press, 1973.

J. Hirschmeier and T. Yui. *The Development of Japanese Business 1600-1980*. London: George Allen & Unwin, 1981.

Japan, Ministry of International Trade and Industry. *Gijutsu Doko Chosa Hokokusho* (Report on the Trend of Technology). Tokyo, 1963.

——. *1982 Tsusho Hakusho* (White Paper on International Trade). Tokyo, 1982.

——. *Wagakuni Sangyo no Gijutsu Suijun narabini Gijutsu Kaihatsuryoku Suijun ni kansuru Ankeito Chosa* (A Questionnaire Survey on Japan's Technological Level and Technological Development Capacity). Tokyo, 1982.

——, Industrial Structure Council. *Hachijunendai no Sangyokozo no Tembo to Kadai* (The Prospects for the Industrial Structure of the 1980s). Tokyo, 1981.

Japan, Ministry of Labor. *1982 Rodo Hakusho* (White Paper on Labor). Tokyo, 1982.

Japan, Science and Technology Agency. *1976 Indicators of Science and Technology*. Tokyo, 1976.

——. *1976 Kagaku Gijutsu Hakusho* (White Paper on Science and Technology). Tokyo, 1976.

——. *1980 Kagaku Gijutsu Hakusho* (White Paper on Science and Technology). Tokyo, 1980.

——. *1982 Indicators of Science and Technology*. Tokyo, 1982.

K. Kojima. *Direct Foreign Investment. A Japanese Model of Multinational Business Operations*. London: Croom Helm, 1978.

S. B. Levine and H. Kawada. *Human Resources in Japanese Industrial Development*. Princeton, N.J.: Princeton University Press, 1980.

L. H. Lynn. *How Japan Innovates. A Comparison with the U.S. in the Case of Oxygen Steelmaking*. Boulder, Colo.: Westview Press, 1982.

N. Lyons. *The Sony Vision*. New York: Crown Publishers, 1976.

M. Morishima. *Why has Japan "Succeeded"? Western Technology and the Japanese Ethos*. Cambridge: Cambridge University Press, 1982.

T. Nakamura. *The Postwar Japanese Economy*. Tokyo: University of Tokyo Press, 1981.

Y. Ojimi. "Basic Philosophy and Objectives of Japanese Industrial Policy." In *The Industrial Policy of Japan*, edited by Organization for Economic Cooperation and Development. Paris: OECD, 1972.

T. Oshima. *Jidosha Sangyo* (The Automobile Industry). Tokyo: Toyo Keizai, 1980.

T. Ozawa. "Technology Transfer and Japanese Economic Growth in the Postwar Period." In *Technology Transfer and Economic Development*, edited by R. Hawkins and A. J. Prasad. Greenwich, Conn.: JAI Press, 1981.

———. "Government Control over Technology Acquisition and Firms' Entry into New Sectors: The Experience of Japan's Synthetic-Fibre Industry." *Cambridge Journal of Economics* 4 (June 1980): 133-46.

———. *Multinationalism, Japanese Style. The Political Economy of Outward Dependency*. Princeton, N.J.: Princeton University Press, 1979.

———. *Japan's Technological Challenge to the West, 1950-1974: Motivation and Accomplishment*. Cambridge, Mass.: MIT Press, 1974.

M. J. Peck and Shuji Tamura. "Technology." In *Asia's New Giant: How the Japanese Economy Works*, edited by H. Patrick and H. Rosovsky. Washington, D.C.: Brookings Institution, 1976.

N. Rosenberg. *Inside the Black Box: Technology and Economics*. Cambridge: Cambridge University Press, 1982.

Royal Academy of Arts. *The Great Japan Exhibition. Art of the Edo Period 1600-1868*. London, 1981-82.

Nikkan Kogyo Shimbun. *Roboto Sangyo Chizu* (Robot Industrial Map). Tokyo: Nikkan Kogyo Shimbun, 1982.

8

Technology Transfer to Japan: What We Know, What We Need to Know, and What We Know That May Not be So

Leonard H. Lynn

INTRODUCTION

Japan has frequently been characterized as the best of all nations at borrowing foreign technology.[1] The Japanese are seen as carefully scanning the world for new technology, adroitly controlling the conditions of purchase (or even pressuring reluctant foreign firms to share it with Japanese firms), and then concentrating their research efforts on incremental improvements that allow them to outpace the inventors in exploiting the technology.

Like all stereotypic images, this characterization has its basis in aspects of reality but also includes assumptions that are not well supported and generalizations that have not been critically evaluated. Beyond this, the base of information is not robust enough to support answers to several questions that are of growing interest, for example: Do the Japanese mechanisms for finding out about foreign technology provide a useful model for other countries? How much, if any, of Japan's success in using foreign technology can be attributed to government policy? What, if anything, should other governments do either to emulate or to cope with these policies? What are the extent and pattern of Japan's dependence on foreign technology? Is this dependence rising or falling? To what extent is Japan emerging as a net exporter of technology? What kinds of technology is it exporting, and to whom?

My purpose here is to review the evidence supporting our understanding of technology transfer to Japan and to indicate its limitations.

I also suggest some areas of research that could help fill in a few of the gaps.

THE JAPANESE SEARCH FOR TECHNOLOGY

Reference is increasingly made to Japan's widespread collection of information on foreign technology. The overseas offices of business firms, trade associations, trading companies, and newspapers, as well as government and semigovernment agencies, all appear to be a part of this in elaborate networks that allow routine attendance at engineering conferences as well as the systematic collection and translation of foreign technical articles and patents. A stream of personnel is sent to foreign universities and on industrial inspection tours. Information is carefully collated and analyzed.[2]

The Japanese have been noted for sending personnel to overseas research organs since early in the Meiji period (1868-1914). Around the turn of the century Japanese engineers studied not only at prestigious universities in the United States and Europe, but also at the research laboratories of such firms as Westinghouse and General Electric.[3] Interestingly, this effort to learn by going overseas appears to have intensified, rather than slackening, as Japan has overtaken Western countries in many areas of technology. In 1978 an estimated 40,000 Japanese made trips to the United States to "study and acquire American technology."[4] About 150 Japanese scientists and engineers are at the Massachusetts Institute of Technology at any given time. In 1981, Pittsburgh's Carnegie-Mellon University had engineering students from Hitachi, Mitsubishi Heavy Industries, Sumitomo Metals, and Nippon Steel Corporation, as well as the Japanese Science and Technology Agency and the Ministry of Construction. At least 50 of Japan's largest business firms have programs in which they send employees to foreign universities to study technology or management. Hitachi sends about 30 a year. Nippon Steel Corporation reports that it sends about four times as many as it did ten years ago.*

---

*The information on students sent by Japanese firms is based on research I am currently carrying out under a grant from the Mellon Foundation Program on Technology and Society.

The collection of technical information by Japanese trading companies is also of long standing. Over half a century ago, Mitsubishi had set up subsidiaries in Europe to identify technologies that might be introduced into Japan. Today each of the half-dozen largest general trading firms has well over 100 overseas offices, and Mitsubishi claims the world's largest private telecommunications network. In 1982, Mitsubishi and Marubeni established technology affairs departments to increase their ability to facilitate the flow of technology.

Numerous other examples could be given of the means by which the Japanese keep track of foreign technical developments. One might mention the Japanese Information Center of Science and Technology, for example. This public corporation was established in 1957 by the Science and Technology Agency. Two-thirds of its funds come from the government, and it collects 10,000 foreign and domestic patents and technical papers each year and has them abstracted by five thousand scientists and engineers.[5]

It seems, then, that the Japanese employ a broad range of activities in their search for technology. Some of these appear to be unique, at least in terms of their intensity. The problem is that we know little about their significance or effectiveness. The Japanese may well have based their economic miracle on an unusual ability quickly to spot promising new technologies developed in other countries (as many Americans would like to believe), but critical questions remain to be answered. It may be worth noting, for example, that some Japanese researchers complain that the intensity with which Japan scans foreign technological developments diverts resources from original research that could lead to better technology than that being borrowed. And even if it were shown that the Japanese gain a large return on their investment in technological surveillance, we do not know enough to conclude that other countries would find it profitable to imitate the Japanese practices. One authority argues that because of the nature of its training, employment, and professional institutions, Japan has had highly constrained flows of technical information. It is partly to compensate for this that the Japanese have invested so heavily in the surveillance of foreign technology. Conversely, he argues, the lack of openness in the Japanese system implies that foreigners would gain little of value by investing heavily in surveying Japanese technology.[6] These assertions remain to be tested.

TECHNOLOGY TRADE AND INDUSTRIAL POLICY

The Japanese Ministry of International Trade and Industry (MITI) has, at times, used its control over technology imports to promote the use of advanced technology by Japanese firms, to improve the bargaining position of Japanese firms dealing with foreign suppliers, to facilitate the diffusion of new technology within Japan, and to shape Japanese industrial structure. This role, however, has changed sharply over the past few decades, causing some confusion between what MITI was able to do in the past and what it is able to do in the present.[7]

Very few agreements to import technology into Japan were made in the first few years after World War II. One problem was that the Anti-monopoly Act of 1947 appeared to ban agreements for the exclusive use of technologies or know-how. This made foreign firms reluctant to sell technologies to Japan. Imports of technology were resumed in 1949 and 1950 as new laws clarified the situation and protected foreign sellers. The new laws also gave sweeping powers to the Japanese government (particularly to MITI) to regulate the import of technology. The Anti-monopoly Act was amended in 1949, and many of the patent licenses suspended during the war were revived. The Foreign Exchange and Foreign Trade Control Law of 1949 gave the government new control over technology imports by requiring that all external transactions involving foreign currency be approved by the government. The Foreign Investment Law of 1950 guaranteed that funds would be available to meet foreign obligations. It also restricted technology imports to those that would improve Japan's balance of payments, develop important industries, or develop public utilities.

These laws were significantly liberalized over the 1960s and 1970s. In 1959 and 1960 there was a shift from a rationale of "positive screening" (allowing only the imports of technology that clearly served the national interest) to "negative screening" (allowing all imports of technology that were not detrimental to the national interest). After its status under the International Monetary Fund changed in 1964, Japan could no longer control its foreign exchange market as tightly. Japan also joined the Organization for Economic Cooperation and Development and in the process agreed to a four-stage plan for capital liberalization. The most important liberalization related to technology imports under this plan occurred on

June 1, 1968. Contracts involving payments of below $50,000 were to be granted automatic approval by the Bank of Japan, except in certain designated industries (for example, aircraft, weapons, explosives, nuclear energy, space, and computers). Liberalization was extended to these designated industries in the 1970s. Procedures were simplified in 1978, and the Foreign Exchange Law was revised in 1979.

To put this in crude summary, one can think of Japanese policy toward technology imports as having been one of rigid control in the 1950s; as having slowly yielded to world pressure for liberalization in the 1960s, but only for industries that were not given high priority by the Japanese; as continuing this process of liberalization in the 1970s; and as having begun to take new, more subtle forms in the 1980s. It is thus crucial that the time period be specified in any discussion of Japanese policies on technology imports.

How did MITI make use of this apparatus in carrying out its industrial policies? Case studies give us some insight for some periods. One such case is that of the basic oxygen steelmaking process.[8] This technology was developed in Austria in the early 1950s. The Japanese signed license agreements for the technology in 1956 and built their first basic oxygen furnace (BOF) in late 1957. By 1962 all of the major integrated Japanese steelmakers were operating BOFs, a milestone that was not reached in the United States for nearly another decade. Other instances that are cited to demonstrate how MITI has intervened in the import of technology to the advantage of Japanese firms include synthetic fiber, petrochemical, computer, and semiconductor technology.[9]

MITI and the Cost of Technology

MITI is frequently credited with (or accused of) helping Japanese buyers get foreign technology at low prices by eliminating competition between potential Japanese buyers. The BOF seems to provide a classic example of this. After two Japanese steelmakers showed signs that they might bid up the price of license rights to the technology, the director of MITI's Iron and Steel Production Division arranged a meeting between the presidents of the firms. It was agreed that only one would negotiate to buy the technology, but that the other would have equal rights to it. As a result Japanese steelmakers were able to get the technology at a per ton cost that was only a small fraction of that paid by steelmakers in other countries.* Intervention by MITI to lower royalties has also been noted in the introduction of other technologies. It may have helped Teijin and Toyo Rayon, for example, to get lower royalty rates from Imperial Chemical Industries (ICI) for the production of Terylene.[10]

MITI and the Rapid Use of New Technology

Aside from helping steelmakers to get the BOF technology at low cost, MITI officials also took other steps to encourage the introduction of the BOF in Japan. The director of the Iron and Steel Production Division wrote some of the first articles published in Japanese about the BOF. MITI continued to promote the technology after the first licensees built their plants, and it was instrumental in the formation of an industry group that exchanged information on the BOF. This facilitated the rapid adaptation of the technology to Japanese conditions, helped the Japanese make several important improvements in it, and made it easier for other steelmakers to use it effectively. As part of the agreement under which one steelmaker was allowed to license the BOF, MITI insisted that all Japanese steelmakers have access to the technology at the same low royalty rate. MITI also helped ensure the availability of funds as part of the steel industry rationalization programs.

Lest MITI's powers be exaggerated, however, it should be pointed out that it failed to persuade Kawasaki Steel to use the BOF in its expansion in the late 1950s, and it later failed to discourage another steelmaker from licensing the ill-fated Kaldo steelmaking process. It was clearly the steel firms that decided what technology they wanted to use.

---

*Not all of this difference can be credited to MITI. Another factor appears to have been that the Austrians (like everyone else) grossly underestimated the growth prospects of the Japanese steel industry.

MITI Pressure on Foreign Firms to Share Technology

A frequent complaint is that MITI has forced foreign firms to share their technology with their Japanese rivals in exchange for the right to do business in Japan. An instance that is often cited to demonstrate this occurred at the beginning of the 1960s when the effort of International Business Machines (IBM) to retain full ownership of its Japanese subsidiary were challenged by MITI. Chalmers Johnson characterizes the IBM case as follows: "IBM ultimately had to come to terms. It sold its patents and accepted MITI's administrative guidance over the number of computers it could market domestically as conditions for manufacturing in Japan."[11] Others speak of IBM "capitulating" to MITI[12] and of the Japanese computer industry owing its origins to the group of patents granted by IBM as part of this arrangement.[13]

Curiously, however, accounts of the incident from the IBM side do not describe it as a MITI victory. One notes:

> The company mounted a three-month task force to survey the problem. Could the Japanese manufacturers make comparable machines? Once he knew the customers would continue to deal with IBM no matter what their government did or said, Gordon Williamson got the backing of Jones and both Watsons to take a hard line with the government negotiators, telling them (most regretfully and politely) that IBM would have to close down its operations rather than yield on the ownership issue. During the survey phase the customers had not only been canvassed about their loyalty, but once that was assured they had been encouraged . . . to bring to the government's attention the dangers to the Japanese economy if IBM should leave.
>
> With its industry support evaporating rapidly, the government asked Williamson to have the IBM board review his decision. When he refused in a final dramatic meeting, the government gracefully withdrew its demands.[14]

How can we reconcile such radical differences of interpretation of what happened? It might be noted that Johnson and the others emphasizing MITI's role in this episode draw heavily on the memoirs of Shigeru Sahashi in reaching their conclusions. Sahashi, a former vice minister who has been called "Mr. MITI," dramatizes the event by saying that: "IBM was the heavyweight champion of the world, while we were in the 'mosquito class.'" He then relates how he gained IBM's respect through his toughness in dealing with them.[15] The IBM account is based on interviews with company executives. Presumably there is much that is self-serving on both sides. A careful and balanced review of the case is needed to allow us to judge how much power MITI was able to exercise.

The Effectiveness of MITI's Use of Its Power

Our doubts about what happened with IBM notwithstanding, it seems clear that MITI has frequently had the power to influence considerably the terms under which technology was transferred to Japan. How much has Japan benefited (or suffered) from the use of this power? Here again, we need to know much more.

MITI's involvement in the introduction of the BOF was a model of expertise brought to bear for the Japanese national interest, but there are indications that MITI was not always so adept in using its power. Trezise and Suzuki say that Sony was delayed for nearly two years in its efforts to import transistor technology because an MITI official felt that it would be unable to exploit the technology adequately.[16] It may be that the level of competence was unusually high in the Iron and Steel Production Division, or perhaps that circumstances were fortuitous in the case of the BOF – this particular case was chosen for study precisely because it was a success. How common were MITI's failures and what problems did they cause? Not enough is known to judge. One case that might fruitfully be studied is Texas Instruments' four-year battle in the mid- and late 1960s to set up a wholly owned subsidiary to make integrated circuits in Japan. This case is frequently cited to demonstrate MITI's power to force foreign firms to give up rights to their technology. MITI's victory, however, may have been Pyrrhic. Trezise and Suzuki quote the chairman of the board of Texas Instruments as speculating that the delays resulting from MITI's interference possibly forced the Japanese to waste time repeating much of the U.S. experience with second-generation computers, thus missing an opportunity to attain parity with or even a lead over the U.S. industry in producing third-generation computers.[17] Since our information about this episode is limited to a few anecdotal paragraphs it is difficult to evaluate.

Evaluating What We Know about the Role of MITI

Much of our "conventional wisdom" about MITI's intervention in Japan's import of foreign technology, then, is based on uncritically accepted anecdotes. Well-documented case studies are scarce, and their scarcity brings the risk that conclusions based on a study of one period or one industry will be inappropriately applied to other periods or other industries. Most of the information we have concerns events in the 1950s and 1960s. The BOF, for example, was introduced in the late 1950s at a time of peak influence for MITI. This was also a period in which steel was among four industries given the highest priority in Japan's economic planning. Priorities changed, and new laws allowed substantially less meddling in foreign agreements. Thus, while Japanese steel production in the late 1970s was 10 to 15 times greater than in the mid-1950s, the Iron and Steel Production Division's staff was one-third smaller.[18] Studies of the role of government in the 1970s and 1980s should also be made.

There may also be rich, but so far neglected, opportunities for other research as well. An examination of government organization charts and biographies of key personnel, for example, could fill many gaps in our understanding of the role of MITI in the 1950s and 1960s, as well as more recently. It is known that some former mid-level MITI people have moved into technical positions in industry, but is not known how often this happens or what significance it has. Knowledge of the kinds of positions they take might give us some sense of the relative position of MITI in its dealings with industry.[19]

TRENDS IN JAPAN'S INTRODUCTION OF TECHNOLOGY

The Japanese government publishes an abundance of statistical data related to R & D and technology transfer. These data give us some sense of the volume of foreign technology introduced in Japan since World War II and are useful in exploring trends. They must, however, be used with considerable caution.

The major source of data on Japanese spending for the purchase of foreign technology is the *Gaikoku Gijutsu Dohnyu Nenji Hohkoku* (Annual Report on the Introduction of Foreign Technology) of the Science and Technology Agency. This report, which I will subsequently refer to as the *Foreign Technology Report*, shows that between fiscal 1950 and fiscal 1980, 35,996 license agreements were concluded to introduce technology into Japan. These agreements cost Japan a total of $11.6 billion. Boretsky, Abegglen and Hout, and others have used this data in comparison with U.S. spending on R & D in an effort to show that the Japanese have saved vast sums of money by buying U.S. and other foreign technology instead of developing their own.[20] A typical statement in this vein is made by James Abegglen: "Between 1950 and 1980, the Japanese essentially acquired all of the technology in the world that they considered worth having for a small fraction of the current annual U.S. expenditure in research and development."[21]

The $11.6 billion that the *Foreign Technology Report* indicates was spent by the Japanese on technology imports was, indeed, only a small fraction of the $61.1 billion the United States spent on R & D in 1980.[22] But, so too, the $6 billion the United States spent on technology imports between 1950 and 1980 was only a fraction of the $19 billion the Japanese spent on R & D in 1979.[23] In short, both the United States and Japan spend far more on R & D than they do on the purchase of foreign technology. While Japan may well have acquired technology at bargain rates because of the involvement of MITI (and, perhaps, the inattention of foreign firms), these data are not relevant to this point.

A more useful analysis of royalties and R & D might calculate the ratio of money spent on foreign technology to that spent on R & D by a country. This could serve as a crude measure of dependence on foreign technology. And, in comparing the ratios for Japan and the United States, we do find that the Japanese ratio of dependence in 1978 (the last year for which comparative data are available) was some five or six times higher than the U.S. ratio. As we might expect, the past two decades have brought some convergence as the Japanese ratio has fallen and the U.S. ratio has increased, suggesting that there have been diminishing returns both to Japan's investments in foreign technology and to the United States' reliance on domestic R & D.*

---

*In 1978 the Japanese paid some 261 billion yen for foreign technology while spending some 3.6 trillion yen on R & D – a ratio of .073. The United States spent some $610 million on foreign technology and $48 billion on R & D – a ratio of .013. The Japanese ratio has been decreasing (from .163 in 1961), while the U.S. ratio has been increasing (from .0062 in 1961).

This analysis might usefully be extended to give insights on comparisons between countries and on trends, but its limitations should be recognized and, to the extent possible, adjustments should be made to overcome conceptual problems with the Japanese data.[24] One problem is that the Japanese records of "technology introductions" include items that are only peripherally, if at all, related to what most of us think of as technology. In fiscal 1981, one-fourth of the 2,076 contracts listed in the *Foreign Technology Report* involved the use of trademarks. While many of these also involved the teaching of production techniques, more than 100 were for nothing more than the right to use trademarks. Some of the "technology introductions" are patents in which no transfer of know-how is involved. A firm may, for example, purchase foreign patent rights to eliminate legal barriers to entry into a market. In a well-known instance of this, Toray purchased rights to DuPont's nylon patents (which also helped Toray establish a monopoly for nylon in Japan).[25] Indeed, in fiscal 1981 some 11 percent of the cases of major technology introduction in Japan were for patents without acquisition of know-how. In the past 15 years or so, many of Japan's technological imports have been for duplicate technology as competing firms sought to protect their positions.

On the other hand, it should not be forgotten that licenses for the use of technology are only one mode of technology transfer. Others include research subcontracting, personnel exchanges, the use of foreign technical publications, attendance at international conferences, overseas training, and industrial espionage.[26] To the extent that the Japanese have been unusually dependent on technology from the United States and other countries, it seems likely that much of it is via some of these other modes of transfer. Yet our knowledge in this area goes little beyond the level of stories businessmen tell over cocktails. More case studies should be developed for use with the quantitative data to give us a firmer basis for evaluating the extent of Japan's reliance on foreign technology.

R & D and Technology Imports

There is a mixture of case and quantitative evidence that the Japanese have invested a substantial part of their R & D spending in the adaptation or improvement of foreign technology. One of the major improvements in the color television picture tube, the black matrix method, was developed by Zenith in the United States. Fully effective exploitation of the technology did not occur, however, until it was improved by the Japanese and incorporated into highly competitive color television sets.[27] This seems to be consistent with our image of the Japanese investing heavily in the modification of foreign technology, improving it slightly, and reexporting it. But how commonly does this really occur? And to what extent do the Japanese pursue "improvement engineering" as a strategy? A set of data that initially appears to give answers to these questions was provided by MITI in a 1963 survey, which showed that about one-third of the R & D spending by the 1,039 firms studied was for the improvement or modification of imported technology. These data have been used in the major analyses of Japanese technology by Ozawa, Peck and Tamura, and Tsurumi.[28]

Aside from the fact that the MITI survey was carried out some 20 years ago, there are other difficulties in evaluating its current relevance. One problem is that we do not have comparable data from other countries – thus we do not have much of a sense of how high 33 percent spending on modification of foreign technology is. Would the European countries have lower ratios? What would the U.S. ratio be? Given the cultural differences between Japan and the sources of most of its imported technology, it seems likely that in many cases (particularly consumer goods) modifications were necessary to accommodate differences in Japanese customs and life-styles. Fewer such modifications would have been necessary, presumably, in the case of technologies being exchanged between Western industrial countries. Be this as it may, it seems to me that another problem is even more serious. Published reports on the survey are not clear as to exactly what was meant by "foreign technology." How, for example, would we count Sony's research on the video tape recorder (VTR)? The VTR was first developed by Ampex, a U.S. company, for commercial broadcast use. Sony developed a VTR for home use that could sell for about one-hundredth the cost of the original Ampex VTR. Should Sony's R & D be counted as part of the research for the improvement and modification of foreign technology?[29] Or should the home VTR be considered a different technology? If one has very strict criteria of "newness," it may be that virtually all R & D done anywhere is for the improvement of foreign technology.

The Balance of Technology Trade

According to *Foreign Technology Report* data, Japan has had a large technology trade deficit since the current system of record-keeping was initiated in 1950 (see Table 8.1). This deficit has increased over the years to reach 1.2 billion dollars in 1981, but Japan's receipts for technology have steadily increased as a percentage of its payments for technology. In the late 1950s receipts only amounted to 1 percent of payments. By the mid-1960s receipts had increased to 10 percent and by the mid-1970s to more than 20 percent of the value of payments. In 1981 receipts came to 31.4 percent of payments. Despite this progress (and Japan's emerging reputation for high technology), Japan in the 1980s was still paying more than three times as much for technology as it was receiving from other countries.

These data, however, include important biases (in addition to the limitations mentioned above). The data for technology exports are less comprehensive than those for imports.* This exaggerates the apparent Japanese technology trade deficit. Another limitation of the annual *Foreign Technology Report* data on technology trade is that they include royalty payments being made for technology agreements concluded in previous years. Since Japanese receipts for technology have been rising much faster than payments, far more royalties for past contracts are paid than received.† Thus, the current

---

*The data on technology imports are collected by the Bank of Japan based on applications for permission to import "foreign technology." Since technology imports are regulated, these data are precise and considered to be of high reliability (though, as was noted above, the definition of "foreign technology" is a very broad one). No procedures exist to control the export of technology. The Bank of Japan assembles its technology export statistics by collecting information on foreign remittances sent to Japan for technology purchases. Transactions such as those for cross-licensing that do not involve remittances are not included. More important to the point here, these data do not include technology embodied in exports of plants and equipment. For an evaluation of the various data sources, see Nippon Sangyo Gijutsu Shinkokai, *Kenkyu Kaihatsu Oyobi Gijutsu Koryu ni Kansuru Chosa Hokokusho, 1973/74* (Tokyo: Nihon Sangyo Gijutsu Shinkokai, 1976), pp. 98-99.

†In 1980 only 12 percent of the 240 billion yen paid by Japanese firms for technology imports were for new agreements. Some 47 percent of the 160 billion yen received for technology exports were for newly concluded contracts. See Japan, Prime Minister's Office, *Kagaku Gijutsu Kenkyu Chosa 1981*, (Tokyo: Nippon Tokei Kyokai, 1982), p. 34.

TABLE 8.1
Japan's Technological Trade 1950-81

F.Y.	*Receipts (U.S. $ millions)* *A*	*Payments (U.S. $ millions)* *B*	*Deficit* *A–B*	*(%)* *A/B*
1950	0.0	2.6	–2.6	–
1951	0.0	6.7	–6.7	–
1952	0.0	9.9	–9.9	–
1953	0.1	13.9	–13.8	0.7
1954	0.4	15.8	–15.4	2.5
1955	0.2	20.0	–19.8	1.0
1956	0.3	33.3	–33.0	0.9
1957	0.2	42.6	–42.4	0.5
1958	0.7	47.8	–47.1	1.5
1959	0.8	61.9	–61.1	1.3
1960	2.3	94.9	–92.6	2.4
1961	3	113	–110	2.7
1962	7	114	–107	6.1
1963	7	136	–129	5.1
1964	15	156	–141	9.6
1965	17	166	–149	10.2
1966	19	192	–173	9.9
1967	27	239	–212	11.3
1968	34	314	–280	10.8
1969	46	368	–322	12.5
1970	59	433	–374	13.6
1971	60	488	–428	12.3
1972	74	572	–498	12.9
1973	88	715	–627	12.3
1974	113	718	–605	15.7
1975	161	712	–551	22.6
1976	173	846	–673	20.4
1977	233	1,027	–794	22.7
1978	274	1,241	–967	22.1
1979	342	1,260	–918	27.1
1980	378	1,439	–1,061	26.3
1981	537	1,711	–1,174	31.4

Source: Japan, Science and Technology Agency, *Gaikoku Gijutsu Donyu Nenji Hohkoku*, Fiscal 1981, p. 49. Original data from Bank of Japan.

balance of payments deficit for technology largely reflects Japan's past rather than its current position as a trader in technology.

Another set of data is sometimes used to overcome this problem. The *Kagaku Gijutsu Kenkyu Chosa Hohkoku* (*Report on the Survey of Research and Development*) of the Statistical Bureau of the Prime Minister's Office, which I will refer to as the *R & D Report*, gives data on technological trade collected from an annual survey of Japanese companies. The survey has separately reported payments for newly concluded contracts for each year since 1972.

As Table 8.2 shows, *R & D Report* data indicate that Japan has had a surplus in its current technology trade balance in every year since 1972. The ratio of receipts to payments has ranged from 1.2 in 1978 to 2.7 in 1980. There has been, with the exception of 1978, a general trend toward an increase in the ratio. This has led some Japanese government agencies to conclude that current Japanese receipts for technology exports have been larger than payments since 1972 – a conclusion also made in the influential article on Japanese

TABLE 8.2
Balance of Current Technology Trade Payments
(million yen)

	Receipts (New Contracts) (a)	*Payments (New Contracts) (b)*	*a/b*
1980	74,263	27,675	2.7
1979	52,079	26,808	1.9
1978	47,110	38,183	1.2
1977	36,284	16,888	2.1
1976	27,030	17,860	1.5
1975	18,876	13,300	1.4
1974	20,101	14,635	1.4
1973	24,718	19,522	1.3
1972	18,206	14,426	1.3
1971	11,109	15,642	.7

Sources: Prime Minister's Office, Japan, *Kagaku Gijutsu Kenkyu Chosa Hohkoku*, 1981, 1975, 1973, 1972.

technology by Merton Peck and Shuji Tamura and in more general works.[30] Thus it would seem that Japan has been a net exporter of technology for more than a decade.

This conclusion may be closer to the mark than the opposite one based on *Foreign Technology Report* data, but it should be noted that the *R & D Report* data also include biases. The most important is that they do not include payments by firms in the service sector, thus excluding such major traders in technology as the trading firms and department stores.[31] Nor do they include dealings in technology by certain public corporations such as the Japan Atomic Energy Research Institute and the Institute of Physical and Chemical Research. Also, the *R & D Report* data are based on pretax receipts for the sale of technology. Since many of the developing countries purchasing technology from Japan set high tax rates on the purchases, this amount is substantially more than the Japanese firms actually receive for their technology.[32] We know very little about the technology trade balance of the service sector firms and public corporations in Japan. Nor do we have data on how much the use of pretax data inflates Japanese receipts for technology exports. In looking at total payments and receipts, however, we find that the *R & D Report* indicates payments for foreign technology that are typically 10 to 20 percent lower than those in the *Foreign Technology Report*.

So is Japan a net exporter of technology? And, if so, when did it reach this status? Work should be done to allow estimates of the significance of the gaps in the *R & D Report* so that we might answer these questions with more confidence. Indeed, efforts to utilize the *R & D Report* better might yield even bigger payoffs. It gives a wealth of information on Japanese technology imports and exports by industry and by nationality of trading partner going back to 1972.

WHAT WE NEED TO KNOW

There has recently been an explosion of interest in Japan and its uses of technology. Unfortunately, there has not been an explosion of solid information in this area. Insights and interpretations that were tentatively advanced to audiences of specialists have been taken over, stripped of their tentativeness, and put out in slick analyses.

As pundits repeat each other our stereotypes take on a new aura of certainty.

While this process may not be an unusual one, in this case it seems to be unusually extreme. And the implications of it are disturbing. One is that poorly grounded images are being used as a basis for our dealing with Japan. This not only leads to demands toward Japan that may be unreasonable, unrealistic, or unnecessary, but also makes it more difficult to resolve trade problems between Japan and other nations. Perhaps even more unsettling, we have a growing popular literature suggesting that we use our (poorly grounded) images of Japan as a basis for reshaping U.S. institutions. Some even suggest the need for a government role, arguing that the executive branch should monitor the civilian flow of technology to Japan and more carefully assess its "overall economic effect on both U.S. competitiveness and national security."[33] A different (but related) danger is that our process of learning may be corrupted. The market for clear (but possibly misleading) generalizations is a tempting one – and it is being filled.

The Japanese government publishes a wealth of data on technology transfer. Ozawa, Peck and Tamura, and others have made good use of these data. We need, however, to know much more about how they are collected and what their limitations are. What effect, for example, have the major changes in administrative procedures since the 1950s had on the definitions and procedures used in collected data on technology imports? How much does the use of pretax data for technology receipts distort our measures of Japan's current balance of technology trade? Do imbalances in the technology trade of service sector firms and public corporations bias *R & D Report* data? Some information on these studies has been published but not yet carefully examined. Other information should be collected through interviews with present and former Japanese government officials. Case studies of individual technology transfers could be matched to government data to help illuminate our interpretations. Further, in some industries such as steel and electronics there appear to be independent data from trade associations that might be used to check and extend the government data. One might look at one of these sectors in depth over time using interviews and documentary evidence to examine the mechanisms used to collect information on technology.

Beyond the need to understand the processes of technology transfer to Japan, however, there is an opportunity. It is possible that no other country offers such a large volume of published government and trade association data along with such an abundance of sources from which to build case studies (for example, company and industry histories and memoirs of business leaders). The potential is great – not only to improve our understanding of processes in Japan, but also to come to a better understanding of international technology transfer in general.

NOTES

1. Merton Peck and Shuji Tamura, "Technology," in Hugh Patrick and Henry Rosovsky (eds.), *Asia's New Giant* (Washington, D.C.: Brookings Institution, 1976), pp. 525-85.

2. Ezra Vogel, *Japan as No. 1* (Cambridge, Mass.: Harvard University Press, 1973), chap. 3.

3. See, for example, Hoshimi Uchida, "Western Big Business and the Adoption of New Technology in Japan: The Electrical Equipment and Chemical Industries 1890-1920," in Akio Okochi and Hoshimi Uchida (eds.), *Development and Diffusion of Technology* (Tokyo: University of Tokyo Press, 1980), pp. 145-73.

4. Robert Ronstadt and Robert J. Kramer, "Getting the Most out of Innovation Abroad," *Harvard Business Review*, March-April 1982, pp. 94-99. The authors estimate that about 5,000 Americans made similar visits to Japan.

5. Robert Gibson and Barbara Kunkel, "Japanese Information Network and Bibliography Control," *Special Libraries*, March 1980, pp. 155-62.

6. Gary R. Saxonhouse, "Japanese High Technology, Government Policy, and Evolving Comparative Advantage in Goods and Services" (Ann Arbor: Department of Economics, University of Michigan, 1983).

7. English-language overviews of this history are in Robert Ozaki, *The Control of Imports and Foreign Capital in Japan* (New York: Praeger Publishers, 1972), pp. 89-101, and Terutomo Ozawa, *Japan's Technological Challenge to the West: 1950-1974* (Cambridge: MIT Press, 1974). Also see Organisation for Economic Cooperation and Development, *Liberalisation of International Capital Movements: Japan* (Paris: OECD, 1968). For events after 1972, see Japan, Science and Technology Agency, *Gaikoku Gijutsu Donyu Nenji Hohkoku 1980* (Tokyo: Ministry of Finance Printing Office, 1981).

8. See Leonard Lynn, *How Japan Innovates: A Comparison with the U.S. in the Case of Oxygen Steelmaking* (Boulder, Colo.: Westview Press, 1982). The following material is drawn particularly from pp. 52-54, 84-85; 102-3; 107-10; 111-12; and 186-87.

9. See Terutomo Ozawa, "Government Control Over Technology Acquisition and Firms' Entry into New Sectors: The Experience of Japan's Synthetic

Fiber Industry," *Cambridge Journal of Economics* 4 (1980), pp. 133-46; Peck and Tamura, "Technology," and Chalmers Johnson, *MITI and the Japanese Miracle* (Stanford, Calif.: Stanford University Press, 1981).

10. See Ozawa, "Government Control."

11. Johnson, *MITI and the Japanese Miracle*, p. 247.

12. Yoshi Tsurumi, *Japanese Business: A Research Guide with Annotated Bibliography* (New York: Praeger, 1978), p. 66.

13. Roy Hofheinz and Kent Calder, *The Eastasia Edge* (New York: Basic Books, 1982), p. 147.

14. Nancy Foy, *The Sun Never Sets on IBM* (New York: William Morrow, 1975), pp. 156-57.

15. Sahashi Shigeru, *Ishoku Kanryo* (Tokyo: Daiyamondo-sha, 1967), p. 216.

16. Philip Trezise and Yukio Suzuki, "Politics, Government, and Economic Growth," in Hugh Patrick and Henry Rosovsky (eds.), *Asia's New Giant* (Washington, D.C.: Brookings Institution, 1976), p. 798. Also note the short reference to this incident in Nick Lyons, *The Sony Vision* (New York: Crown, 1976), pp. 42-43. The Lyons account, however, suggests that the delay could not have been more than one year.

17. See Trezise and Suzuki, "Politics, Government, and Economic Growth," p. 799. For a brief description of the Texas Instruments episode, see John E. Tilton, *International Diffusion of Technology: The Case of Semiconductors* (Washington, D.C.: Brookings Institution, 1971), pp. 146-47.

18. Interview, Mr. Jin Suzuki, Tokyo, April 1978. Mr. Suzuki was Director of the Iron and Steel Production Division from 1975 to 1977; he was a junior member of the section in the 1950s.

19. This is done for nontechnical officials at the vice-ministerial level in MITI in Chalmers Johnson, "The Reemployment of Retired Government Bureaucrats in Japanese Big Business," *Asian Survey*, November 1974, pp. 953-75.

20. See Michael Boretsky, "Trends in U.S. Technology: A Political Economist's View," in Thomas J. Kuchn and Alan L. Porter (eds.), *Science, Technology and National Policy* (Ithaca, N.Y.: Cornell University Press, 1981), pp. 161-88, and James C. Abegglen and Thomas M. Hout, "Facing up to the Trade Gap with Japan," *Foreign Affairs*, Fall 1978, pp. 146-68 and James C. Abegglen, *Business Strategies for Japan* (Tokyo: Sophia University, 1970), p. 121.

21. James Abegglen, "U.S. Japan Technological Exchange in Retrospect, 1946-1981," in Cecil H. Uyehara (ed.), *Technological Exchange: The U.S.-Japanese Experience* (Washington, D.C.: University Press of America Inc., 1982), pp. 1-13.

22. Japan, Science and Technology Agency (STA), *Gaikoku Gijutsu Donyu Nenji Hohkoku: 1980* (Tokyo: STA, 1981), p. 72, and National Science Board, *Science Indicators, 1980* (Washington, D.C.: National Science Foundation, 1981), p. 210.

23. See Mary Frances Teplin, "U.S. International Transactions in Royalties and Fees: Their Relationship to the Transfer of Technology," *Survey of Current Business* 53 (December 1973), pp. 14-18, and Meryl L. Kroner, "U.S.

International Transactions in Royalties and Fees," *Survey of Current Business* 60 (January 1980), pp. 29-35. These reports give data for most of the years between 1956 and 1978; rough extrapolations were made to yield the estimate.

24. For a good discussion of some of the limitations of royalty data see Jack Baranson, "Critique of International Technology Transfer Indicators," in *Papers Commissioned as Background for Science Indicators – 1980*, vol. 1 (Washington, D.C.: National Science Foundation, 1980), and Edwin Mansfield, "International Trade and Trade Flows," in ibid.

25. See Ozawa, "Government Control."

26. Stefan Robock, *The International Technology Transfer Process* (Washington, D.C.: National Academy of Sciences, 1980), pp. 6-7.

27. Jack Baranson, *The Japanese Challenge to U.S. Industry* (Lexington, Mass.: Lexington Books, 1981), p. 169.

28. Ozawa, *Japan's Technological Challenge*, p. 69; Peck and Tamura, "Technology," p. 542; and Yoshi Tsurumi, *Technology Transfer and Foreign Trade* (New York: Arno Books, 1980), pp. 214-15.

29. Baranson, *The Japanese Challenge to U.S. Industry*, p. 171.

30. Peck and Tamura, "Technology," pp. 525-85. See also Hofheinz and Calder, *The Eastasia Edge*, p. 148. Stuart Kirby and Mary Saso, *Japanese Industrial Competition to 1990* (Cambridge, Mass.: Abt Books, 1982), p. 42.

31. In 1976 six of the ten most frequent importers of technology were trading companies, and another was a department store. See Japan, Science and Technology Agency, *Kokusai Gijutsu Teikei Kaigai Gijutsu Donyu Jokyo to Kaisetsu, 1977* (Tokyo: Nippon Kogyo Shimbun, 1977).

32. Japan, Science and Technology Agency, *Kagaku Gijutsu Hakusho, 1981* (Tokyo: Ministry of Finance, 1981), p. 185.

33. Kent Calder, "Technology Transfers, Promise or Peril?" in *U.S.-Japan Relations in the 1980's: Towards Burden Sharing*, 1981-82 annual report of the Program on U.S.-Japan Relations, Center for International Affairs, Harvard University, pp. 49-58.

REFERENCES

J. Abegglen. *Business Strategies for Japan*. Tokyo: Sophia University, 1970.

——. "U.S. Japan Technological Exchange in Retrospect, 1946-1981." In *Technological Exchange: The U.S.-Japanese Experience*, ed. C. Uyehara. Washington, D.C.: University Press of America Inc., 1982.

J. Abegglen and T. M. Hout. "Facing up to the Trade Gap with Japan." *Foreign Affairs* 57 (Fall, 1978): pp. 146-68.

J. Baranson. "Critique of International Technology Transfer Indicators." In *Papers Commissioned as Background for Science Indicators – 1980*, vol. 1. Washington, D.C.: National Science Foundation, 1980.

——. *The Japanese Challenge to U.S. Industry*. Lexington, Mass.: Lexington Books, 1981.

M. Boretsky. "Trends in U.S. Technology: A Political Economist's View." In *Science, Technology and National Policy*, ed. T. J. Kuehn and A. L. Porter. Ithaca, N.Y.: Cornell University Press, 1981.

K. Calder. "Technology Transfers, Promise or Peril?" In *U.S.-Japan Relations in the 1980's: Towards Burden Sharing*, 1981-82 annual report of the Program on U.S.-Japan Relations, Center for International Affairs, Harvard University.

N. Foy. *The Sun Never Sets on IBM*. New York: William Morrow, 1975.

R. Gibson and B. Kunkel. "Japanese Information Network and Bibliography Control." *Special Libraries* (March, 1980): pp. 155-62.

R. Hofheinz and K. Calder. *The Eastasia Edge*. New York: Basic Books, 1982.

Japan, Prime Minister's Office. *Kagaku Gijutsu Kenkyu Chosa Hohkoku 1981*. Tokyo: Nippon Tokei Kyokai, 1982.

Japan, Science and Technology Agency. *Gaikoku Gijutsu Donyu Nenji Hohkoku: 1980*. Tokyo: STA, 1981.

——. *Gaikoku Gijutsu Donyu Nenji Hohkoku 1980*. Tokyo: Ministry of Finance Printing Office, 1981.

——. *Kagaku Gijutsu Hakusho: 1981*. Tokyo: Ministry of Finance, 1981.

——. *Kokusai Gijutsu Teikei Kaigai Gijutsu Donyu Jokyo to Kaisetsu, 1977*. Tokyo: Nippon Kogyo Shimbun, 1977.

C. Johnson. *MITI and the Japanese Miracle*. Stanford, Calif.: Stanford, University Press, 1981.

——. "The Reemployment of Retired Government Bureaucrats in Japanese Big Business." *Asian Survey* 14 (November, 1974): pp. 953-75.

S. Kirby and M. Saso. *Japanese Industrial Competition to 1990*. Cambridge, Mass.: Abt Books, 1982.

M. L. Kroner. "U.S. International Transactions in Royalties and Fees." *Survey of Current Business* 60 (January, 1980): pp. 29-35.

L. Lynn. *How Japan Innovates: A Comparison with the U.S. in the Case of Oxygen Steelmaking*. Boulder, Colo.: Westview Press, 1982.

N. Lyons. *The Sony Vision*. New York: Crown, 1976.

E. Mansfield. "International Trade and Trade Flows." In *Papers Commissioned as Background for Science Indicators – 1980*, vol. 1. Washington, D.C.: National Science Foundation, 1980.

National Science Board, *Science Indicators, 1980*. Washington, D.C.: National Science Foundation, 1981.

Nippon Sangyo Gijutsu Shinkokai, *Kenkyu Kaihatsu Oyobi Gijutsu Koryu ni Kansuru Chosa Hokokusho, 1973/74*. Tokyo: Nihon Sangyo Gijutsu Shinkokai, 1976.

Organization for Economic Cooperation and Development. *Liberalisation of International Capital Movements: Japan*. Paris: OECD, 1968.

R. Ozaki. *The Control of Imports and Foreign Capital in Japan*. New York: Praeger Publishers, 1972.

T. Ozawa. "Government Control over Technology Acquisitions and Firms' Entry into New Sectors: The Experience of Japan's Synthetic Fiber Industry." *Cambridge Journal of Economics* 4 (1980): pp. 133-46.

——. *Japan's Technological Challenge to the West: 1950-1974*. Cambridge, Mass.: MIT Press, 1974.

M. Peck and S. Tamura. "Technology." In *Asia's New Giant*, ed. H. Patrick and H. Rosovsky. Washington, D.C.: Brookings Institution, 1976.

S. Robock. *The International Technology Transfer Process*. Washington, D.C.: National Academy of Sciences, 1980.

R. Ronstadt and R. J. Kramer. "Getting the Most Out of Innovation Abroad." *Harvard Business Review* 60 (March-April, 1982): pp. 94-99.

G. R. Saxonhouse. "Japanese High Technology, Government Policy, and Evolving Comparative Advantage in Goods and Services." Ann Arbor: Department of Economics, University of Michigan, 1983.

S. Shigeru. *Ishoku Kanryo*. Tokyo: Daiyamondo-sha, 1967.

M. F. Teplin. "U.S. International Transactions in Royalties and Fees: Their Relationship to the Transfer of Technology." *Survey of Current Business* 53 (December, 1973): pp. 14-18.

J. E. Tilton. *International Diffusion of Technology: The Case of Semiconductors*. Washington, D.C.: Brookings Institution, 1971.

P. Trezise and Y. Suzuki. "Politics, Government, and Economic Growth." In *Asia's New Giant*, ed. H. Patrick and H. Rosovsky. Washington, D.C.: Brookings Institution, 1976.

Y. Tsurumi, *Japanese Business: A Research Guide with Annotated Bibliography*. New York: Praeger, 1978.

———. *Technology Transfer and Foreign Trade*. New York: Arno Books, 1980.

H. Uchida. "Western Big Business and the Adoption of New Technology in Japan: The Electrical Equipment and Chemical Industries 1890-1920." In *Development and Diffusion of Technology*, ed. A. Okochi and H. Uchida. Tokyo: University of Tokyo Press, 1980.

E. Vogel. *Japan as No. 1*. Cambridge, Mass.: Harvard University Press, 1973.

9

Licensing Versus Foreign Direct Investment in U.S. Corporate Strategy: An Analysis of Aggregate U.S. Data

Farok J. Contractor

INTRODUCTION

This chapter addresses the question, How important is licensing in the foreign operations of U.S. firms? A majority of agreements are of course between the U.S. company and its own controlled foreign affiliate as licensee. In such a case the agreement exists for tax reasons, or for strategic purposes subordinate and complementary to the direct investment. Here the principal focus is on licensing as a complete or partial substitute for direct investment. This chapter first shows that income from licensing to independent and minority joint-venture licensees comprises a larger fraction of total repatriated foreign income than is commonly realized. Second, it examines the current state of theory on how a firm chooses between the licensing and foreign investment options. Third, it examines in a statistical model the foreign country and industry characteristics that influence the choice between licensing and direct investment.

LICENSING AS A SUBSTITUTE FOR OR COMPLEMENT TO INVESTMENT

In strategy, arm's-length licensing plays a variety of roles, ranging from that of an incidental income source to that of the basic means of overseas expansion in some companies. To some, licensing income is akin to "found money," implying that the firm occasionally stumbles upon licensing opportunities abroad but does not actively seek them.* To others, it is the dominant, if not the only viable international strategy. For instance, for a medium-size firm in a mature industry with many global licensees dependent on its technical services, international brand name recognition, market network, and components may provide a relatively low-risk, steady income stream, not only from royalties but also from technical fees and margins on components supplied. This may be preferable to the risks and competition it would encounter in direct investments overseas in its industry. Besides, such firms may not have sufficient managerial and financial resources to make equity investments in the large numbers of countries their product is sold in. This constraint also applies to large conglomerates, for whom the thousands of possible country/product combinations cannot possibly all be exploited by direct investment. In such cases, across-the-board insistence on equity investment may mean a neglect of salable technologies and a failure to exploit incremental income opportunities via licensing.

There is some evidence that this is being recognized with the creation and expansion of licensing departments in many companies. Perhaps more importantly, the inventorying of technical assets and the possibility of licensing as a foreign market entry method are increasingly factored into corporate strategy in a formal fashion. Externally, there is emerging an international technology market serviced by brokers, consultants, and technology data banks, as well as the conventional industry associations and licensing executives' societies. Last, most observers agree that the international business environment has become somewhat more stringent for the traditional methods of expanding overseas, namely equity investment or exporting, compared with the less constrained postwar period, which ended in the early 1970s. Thus far there is, however, no evidence in the aggregate data of a shift away from direct investment.

Given the traditional and continuing preference for direct equity investment as the desired method for entering overseas markets, at least among the larger and higher-technology U.S. firms, licensing already contributes a surprisingly large fraction of foreign income derived by U.S. companies as a whole. For instance, in 1979, net

---

*Unless otherwise indicated the word *licensing* in this chapter refers to its use as a primary strategy, as opposed to its secondary role as an adjunct or afterthought to direct investment.

nsing fees and royalties totaled about $7.3 billion compared with
5 billion in net foreign affiliate dividend income. The place of licensing
and royalties in total foreign income and their strategic role remain
ificant even after we account for the fact that a majority of licensing
urs between a U.S. company and its own foreign affiliates. About $2.5
on might be described as foreign licensing income from independent
nsees and those in which the U.S. firm has less than a controlling inter-
leaving $4.8 billion in net licensing income from majority foreign
sidiaries.

Why would a firm sign a licensing agreement with its own controlled
liate? When there is an equity stake, the repatriation of declared divi-
ds of the foreign affiliate might appear to be sufficient means for
action of a return to the parent firm, until we realize how many con-
ints exist on this income channel. Licensing then not only provides an
iliary channel but has other important taxation and strategy ramifica-
s. Royalties and fees may be set at the internal discretion of the multi-
onal firm and not necessarily reflect the value of the technology trans-
ed (as is more likely to be the case with arm's-length parties). The actual
pensity of U.S. firms to use licensing agreements with affiliates as a
sfer-pricing device is very much an open empirical question, especially
that more governments disallow the deductibility of royalties and
nology fees paid by a company to its own parent.[1] However Kopits[2]
find a tax-induced bias in affiliate royalty determination. But even if
foreign tax treatment of royalties and dividends were, in effect, iden-
l, there remain several important reasons for having a licensing agree-
nt with one's affiliate. First the U.S. Internal Revenue Service requires
n part, because it wishes to tax adequately the foreign fruits of R & D
lars spent in the United States. Tax law is not the purview of this
pter, but we may note in passing that under select conditions the
atriation of foreign income declared as royalties and licensing fees,
considered a return of capital, will attract a lower U.S. tax rate than
dends.[3] A second reason for earning both royalties and dividends from
oreign affiliate is the pragmatic matter of spreading the risk. The firm
etter insulated against ordinary commercial and the so-called political
. Royalties being usually keyed to output, they are stable even when
fits are zero, enabling some return from the business. Also, when day-
lay operations are in the hands of a local partner, royalties are much
re easily audited and monitored than profits. As far as political risk is
cerned, in the event of a foreign exchange crisis and controls on con-
sion of local currency, governments may accord a somewhat higher

priority to requests to repatriate royalties, being contractual commitme
than to dividends, which may remain frozen. In the case of a nationa
tion or expropriation, there have been instances where the new owr
or the government, have offered to continue the licensing arrangement e
when the equity of the enterprise has been taken over.

The principal focus in this chapter, however, is on licensing as a prin
international strategy alternative to direct investment rather than a
adjunct or afterthought to an equity investment decision. Its princ
focus will be on "licensing out," that is, leasing or selling proprietary in
gible assets for income, rather than "licensing in," or acquiring technol
which is a distinct strategy issue.

LICENSING IN INTERNATIONAL OPERATIONS: A REVIEW OF THEORY

Let us initially assume that for a technology a firm owns, internati
markets or countries are segmented. If so, the strategic objective is to m
mize profit in each country. Let us further assume that because of tr
port cost, tariff, or other barriers, production has to occur within a cou
and that the choice is between establishing a fully owned and contro
equity affiliate and arm's-length licensing. In industrial economics te
nology, the choice is between "internalization" of the technology tran
within the international firm or transferring the technology to an inde
dent licensee under the aegis of a transaction or agreement. (A joint-eq
venture represents an intermediate position between the two extremes.)

Figure 9.1 shows a schematic representation of a proposed inte
tional technology transfer, under either licensing or direct investm
For simplicity I assume that two categories of costs C_1 and C_2 are incu
entirely in the technology supplier's country. C_1 is the present valu
sunk research and development costs of this particular technology,
I will make the further assumption that C_1 is the same regardless of
strategy used to commercialize it overseas. C_2 is the present value of
eral administrative headquarters overheads for all technologies and p
ucts (including the cost of failed R & D efforts).

T_1 is the present value of the costs of transferring the technolog
the other nation. This cost, as we shall see, is very much a functior
the strategy used. Similarly, T_2, the present value of actually repatri
earnings or returns on the technology from the nation, is indeed depen
on the institutional arrangements, whether received as licensing fees ur

FIGURE 9.1
Schematic of International Technology Transfer Costs and Compensation in Either Licensing or Direct Investment

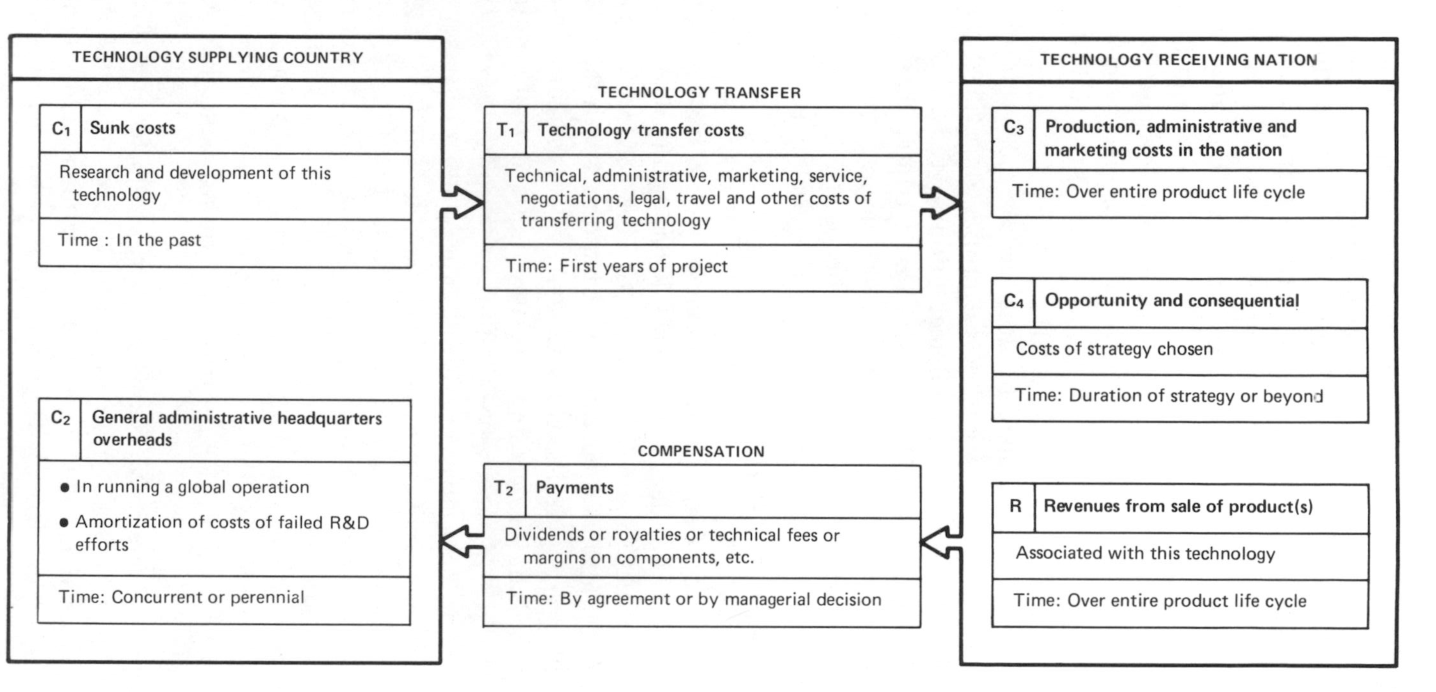

agreement constraints or received as repatriated dividends from an equity investment. Where C_1 may be described as the cost of producing the technology in the first instance, T_1 is the cost of reproducing it in the foreign country, that is, the variable cost of the technology.

Once the technology is transferred, a product is manufactured and marketed at cost C_3, to earn revenues R in the nation. The present value of $R - C_3$ over the product life cycle in the country may be described as the gross margin on the technology from that nation. (Even for the same quantity of product sold each year, neither R nor C_3 will be the same under direct investment and licensing, a point elaborated below.)

Last we have what may be described as the opportunity costs of either strategy. These are the lost profits (or savings) that would have accrued but for the implementation of the strategy, such as the profits on future direct sales to that territory that will now be eliminated. With licensing there may also be so-called consequential costs. For instance, on expiry of the license, or even before, a licensee may compete with the technology supplier in third countries, causing loss of profits there.

It is axiomatic that $T_2 \leqslant R - C_3$; that is to say, licensing fees or dividends repatriated have to come from the gross margin on the technology. But in deciding between licensing and direct investment the focus has to be on $T_2 - T_1 - C_4$, the net technology margin from the country. A focus only on the net direct cash flow $(T_2 - T_1)$ can lead to wrong decisions, particularly in companies with an international division or licensing department as a separate profit center to which $T_2 - T_1$ accrues, while the opportunity costs C_4 are borne by other product divisions. Ideally the company wishes to have $T_2 \geqslant (T_1 + C_4) + \gamma_i C_1 + \gamma_{ij} C_2$ where γ_i is the share of this product's sales in nation i out of global sales of the product ($i = 1 \ldots n$ nations the product is sold in) and where γ_{ij} is the share of the product's sales in nation i out of global sales of all the firm's products ($j = 1 \ldots p$ products). In short, the net direct cash flow $(T_2 - T_1)$ should not only cover opportunity costs of the strategy used, C_4, but also make a partial contribution toward central R & D and overhead.

A priori, one may ask why the risk-adjusted net technology margin $T_2 - T_1 - C_4$ from a licensing agreement cannot be as high as or higher than that from direct investment. Indeed, in several instances, it must. Otherwise we would not have arm's-length licensing.

However, data and theory suggest that it would be the preferred choice in only a minority of situations. Why? There are three related strands in the argument. The first strand of theory deals with the cost of transferring the technology, T_1. In most cases, Casson[4] argues, this will be higher in licensing than an internal transfer to an equity affiliate. Technical information is akin to a public good in the sense that its use in one location does not diminish its stock or availability in another. But its transfer is not costless. Technology transfer is not the transfer of only codified information in patents, manuals, blueprints, and so on. Rather it is the transfer of a capability that may require interpersonal contacts of a long duration. As Teece (1977) or Contractor[5] measured them, international transfer costs are far from trivial. Moreover, they are likely to be higher in licensing because of the technological inferiority of the licensee and the latter's unfamiliarity with the technical or administrative standards and procedures. To these must be added significant negotiations costs in licensing because of what Teece[6] describes as "informational asymmetries." The licensee as the "less-informed party" has to be educated as to the value of the technology without, paradoxically, too much being revealed. This is usually a protracted and costly affair for the prospective licensor. Yet another element of transfer cost that is not present in transactions with an equity affiliate is the legal cost of the license agreement, and patent filing and enforcement, if any.

In brief, transfer costs T_1 in licensing begin to approach the transfer costs of direct investment only when the technology is standardized, codified, and easy to assimilate. Transfer costs are also lower if the recipient company is technologically on a par with the supplier firm.

The second strand of theory deals with the appropriability or extraction of returns from the foreign country. The focus here is on T_2, under licensing and direct investment. A licensor's problem is to try and restrict the licensee's *share* of local revenues to a normal return on entrepreneurial investment, while appropriating the profit on the technology for itself. Even in a fully owned subsidiary the repatriated dividends are only a fraction of profits, that is to say $T_2 \leqslant R - C_3$, but there is no problem of negotiating with an arm's-length party, which can be a vexing affair. Moreover, a prospective licensee usually places a higher uncertainty, and lower value, on the technology because of relative (or in some cases absolute) ignorance, so as to offer license fees below the repatriable profit that may be earned by an equity investment. Dividends, unlike licensing payments, are not constrained by an agreement formula, nor are they of limited duration. Last, Casson[7] suggests another constraint on licensing compensation, namely the possibility that licensing the same technology in several countries puts pressure on the licensor to move to a uniform price for all licensees, which is suboptimal compared to the discriminatory compensation possible in a global equity investment strategy.

The third strand of theory deals with the so-called monopoly power of the international firm in product markets overseas. The focus here is on the product revenues R. A venerable body of literature going back to Caves and Hymer[8] has argued that the international firm is able to command higher prices compared to local competition (or in my argument, compared to a local licensee making an equivalent item). A variety of reasons are enumerated, such as superior product quality, xenophilia, superior organization, international brand recognition, and so on. These need not be examined here; but the point is, for the same production and distribution costs C_3, revenues R may well be higher in one's own subsidiary than those commanded by a licensee. The margin $(R - C_3)$ earned from the product market is likely to be higher therefore in direct investment compared to that earned by a licensee.

To summarize, the net technology margin extracted from a country $(T_2 - T_1 - C_4)$ is held by theory to be more often than not lower in licensing than from a company's own subsidiary because returns T_2 from licensing are likely to be lower while transfer costs T_1 and opportunity costs C_4 are usually higher. Licensing may provide an equivalent or superior margin under select conditions – where the technology is well defined, simple, mature, or standardized, where the licensee is technologically proficient or on a par with the licensor, and where the opportunity costs of licensing are low or zero, such as in developing countries or socialist markets. This is the focus of the empirical analysis.

THE DATA

The objective of this section is to demonstrate that the contribution of licensing revenues to the balance of payments and to company income is considerably larger than commonly assumed, and that in these revenues the share of independent or less than fully controlled licensees is greater than immediately apparent.

The Commerce Department's definition of direct investment in its benchmark 1977 survey is ownership by a single person, association of persons, corporation, or corporate group of at least 10 percent of a foreign business enterprise that is deemed an affiliate. A less than 10 percent interest is not considered direct investment and "is not considered to have sufficient ownership to influence management."[9] While the 10 percent cutoff is as good as any in that range, it is clear that the definition of foreign affiliation is thus very broad and includes minority joint ventures where the U.S. firm has between 10 and 50 percent of equity even though in the latter case the relationship may effectively be at arm's-length. There is only an approximate correlation between the percentage of shares held and the degree of control or influence wielded by the minority partner. To varying degrees, licensing agreements with foreign minority joint ventures tend toward, or may even approximate, an arm's-length relationship over compensation and other agreement terms, because the obvious fact is that while the dividends are to be shared, the licensor gets to keep all of the royalties and fees. In short, the Commerce Department's classification of "affiliate licensing" and "non-affiliate licensing" tends to understate the importance of licensing as a primary international strategy, as an income generator, or as a market-entry method (as opposed to its secondary role in majority affiliates for tax and other strategy considerations).

This immediately poses the question of just what is the breakdown of majority versus minority affiliates, the extent of licensing in each category, and the extent of licensing in the so-called unaffiliated category. The latest figures, released in 1981, come from the 1977 benchmark survey and are reinterpreted and summarized in Tables 9.1-9.9. Interpretation is slightly complicated by the fact that there are two statistical universes of U.S. licensors, one on unaffiliated licensing (with data gathered by the Balance of Payments Division on form BE-93) and the other on licensing with affiliates (with data gathered by the International Investment Division on form BE-577). The 1977 benchmark survey on U.S. direct investment involves an overlap, including in some tables the nonaffiliate licensing of U.S. firms that also have other foreign affiliates.[10] This survey did not include the subset of U.S. licensors that have no affiliates abroad, and their number is unknown, as is shown in Table 9.1. One suspects that much of foreign nonaffiliate licensing is done by firms that also have at least one other foreign affiliate. There is simply no information on this, however. Finally, we may note that there are reasons to suspect serious underreporting in the unaffiliated licensing category which tends to understate its significance.

Figure 9.2 shows 3,540 nonbank U.S. companies with 23,641 nonbank affiliates. (In technology transfer or licensing terms the share of bank parents or bank affiliates is negligible.) Over a thousand U.S. companies had licensing agreements directly with unaffiliated foreigners; 905 of the U.S. firms (also had another foreign affiliate and) account for arm's-length agreements with at least 22,000 foreign companies. In addition there was an indirect transfer of U.S. technology to approximately 7,000 unaffiliated companies under agreements made by foreign affiliates with other foreign parties.

Figure 9.3 examines licensing by U.S. parents to their foreign affiliates and the affiliates' licensing in turn to other foreigners. In Figure 9.2 we saw that, already in 1977, the parents of minority affiliates exceeded the number of parents of majority affiliates. Counting the number of affiliates, there is an almost even division of the total between 11,909 majority and 11,732 minority affiliates. I should quickly add, however, that this can mislead. By assets, or number of employees, the share of minority affiliates was much smaller, as shown in Figure 9.3. Of the 23,641 affiliates, 9,708 received technology or patent and trademark rights from parents. Technology and rights were licensed by 1,701 affiliates to about 15,000 other foreigners. Of these, half were unaffiliated parties. Since licensees abroad receive technology both directly from U.S. firms and indirectly from their foreign affiliates, this means that a small portion of the licensing income on U.S. technology is not reported as such. It accrues to the foreign affiliates, is merged with their profits, and may be then repatriated to the U.S. as affiliate dividends.

Next, let us examine the understatement that often arises because licensing dollar income figures are reported on a net basis, that is to say, gross receipts less payments by U.S. parents to their affiliates for other technology received. Before addressing this issue,

FIGURE 9.2
U.S. Companies' Overseas Affiliates and Licensing, 1977

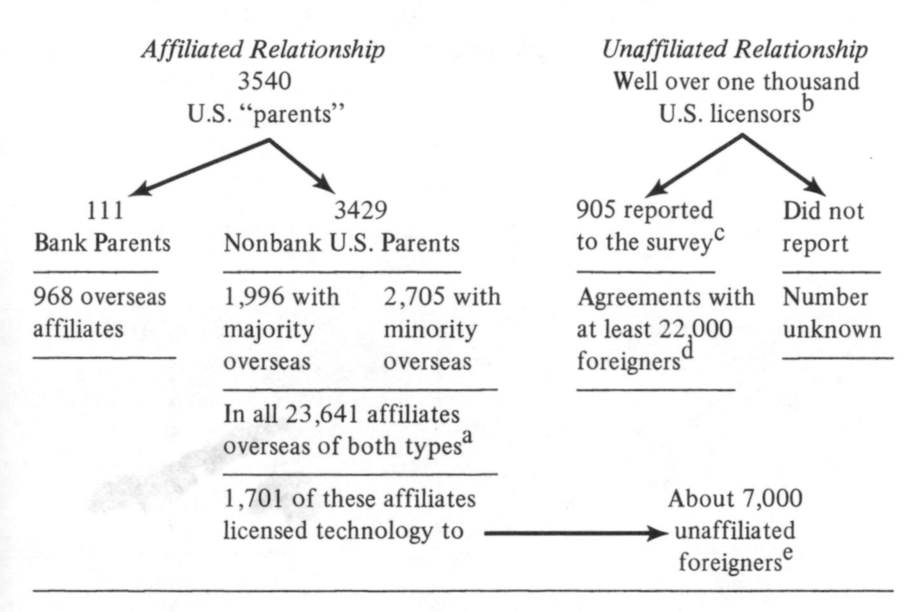

Notes:

[a]The actual figure is 23,698 affiliates but 57 Bank affiliates were removed. From the survey it appears that no "Royalties and License Fees" are paid by bank affiliates at all. They did, however, pay $190 million in "Service Charges and Rentals," almost all of it to their bank parents. (See pages 20, 95 and 217 of the Survey.)

[b]The Balance of Payments Division gets reports from only 650-700 reporters. There are reasons to fear there may be a significant nonreporting problem.

[c]Of the universe of well over a thousand U.S. licensors having unaffiliated licensing agreements abroad, who report to the Balance of Payments Division, 905 reported under this survey to the International Investment Division because they had some affiliates overseas as well. (See page 191 of the Survey.)

[d]The estimate is derived thus: The survey indicates a count of 22,878 for the number of unaffiliated persons with whom there are patent and know-how agreements; this should be one-third higher if trademark agreements were included (see page 191 of the Survey). On the other hand, we should also deflate by about 30 percent to account for the fact that page 191 figures are gross totals for technology transfers both to and from the U.S. company. (For this see page 190.) Thus about 22,000 is a reasonable estimate for the number of unaffiliated companies with which the U.S. firm is a licensor.

[e]Please see Figure 9.3.

Source: U.S. Department of Commerce, *U.S. Direct Investment Abroad, 1977* (Washington, D.C., 1981).

FIGURE 9.3
Licensing to Affiliates and Their Licensing to Others (nonbank affiliates of nonbank parents, 1977)

	Total	*Majority*	*Minority*	*Minority as % of Total*
No. of affiliates[a]	23,641	11,909	11,732	50%
Assets, $ million[a]	490,178	352,257	137,821	28%
Employees[a]	7,196,691	5,368,826	1,827,865	25%

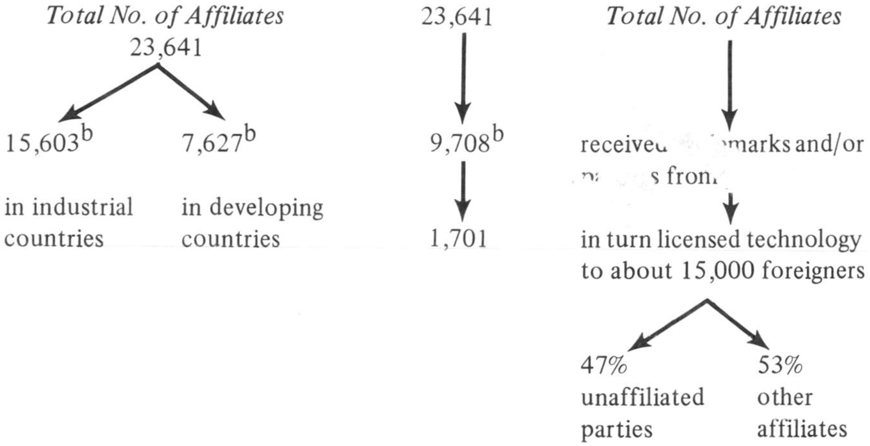

Total No. of Affiliates
23,641
↙ ↘
15,603[b] in industrial countries
7,627[b] in developing countries

23,641
↓
9,708[b]
↓
1,701

Total No. of Affiliates
↓
received [illegible] marks and/or [illegible] s from
↓
in turn licensed technology to about 15,000 foreigners
↙ ↘
47% unaffiliated parties
53% other affiliates

Notes:
[a]Page 20 of the Survey.
[b]Page 162 of the Survey.
[c]10,121 foreigners received patents and process technology in turn from the affiliates. Also, 9,526 foreigners received trademark rights from the affiliates. But since the two subsets overlap, i.e., have a partial union, the total number of foreigners is not the sum of the two, but less, perhaps 15,000 or so, at most. See pp. 166-67 of the Survey.

Source: *U.S. Direct Investment Abroad, 1977.*

one needs to understand the Commerce Department's terminology, which can be confusing. Total gross receipts are broken down into various categories in Table 9.1. *Fees and Royalties* is the term used to describe the grand total of all sums associated with technology transfer. This in turn is broken down into two broad divisions: 1) "Royalties and License Fees" for intellectual property rights transferred to the licensee, such as patents, know-how, and trademarks; and 2) "Service Charges and Rentals" for services to enable the technology transfer, such as technical, managerial, marketing, or research and development assistance. Thus the principal conceptual distinction is between rights and services. This quite correctly recognizes that technology transfer is not merely an act of transferring proprietary information and rights to the other firm, but that attendant services have to be provided to facilitate and effectuate the transfer.[11] For historical reasons, two small payments extraneous to technology transfer are thrown into the pot, namely "Rent of Tangible Property" and "Film and TV Tape Rentals." These are, however, a tiny fraction of the total and ought to be subtracted.

Receipts of fees and royalties from unaffiliated licensees (which of course cannot be netted against affiliates) are reported on a gross basis.* These, added to fees and royalties from affiliates, gives the grand total of receipts of fees and royalties for technology directly licensed from the United States, amounting in 1977 to $5,848 million. (To this, if one wishes, can be added the $295 million received by affiliates licensing technology to yet other unaffiliated foreigners, as shown in Table 9.5.)

Table 9.2 addresses the relationship between gross receipts and receipts netted against affiliates. The essential point is that netting deflated gross receipts by 20 percent in 1977.† Accordingly the reported net figures in Table 9.4 for the years 1972-81 are inflated

*In some publications they are occasionally netted against payments made by U.S. firms *to* some other unaffiliated foreign firms for technology "licensed in."

†The same is not true under "Royalties and License Fees"

	Gross Receipts	*Payment to Affiliates*	*Net Receipts*
All parents	2225	51	2173
Nonbank	2225	51	2173
Minority nonbank	1962	48	1914

TABLE 9.1
Gross Receipts of All Fees and Royalties Directly by U.S. Licensors, 1977
($ millions)

	Breakdown of Affiliate Fees and Royalties							
		Service Charges & Rentals						
	Royalties and License Fees	*Services Rendered*	*R & D Assessment*	*Rentals for Tangible Property*	*Film & Tape Rental*	*Total Affiliate Fees and Royalties*	*Total Nonaffiliate Fees and Royalties*	*Grand Total Fees and Royalties*
All U.S. parents[b]	2,225	1,931	227	174	300	4,857	991[a]	5,848
Nonbank parents/ nonbank affiliates[c]	2,225		(2,143) Total		300	4,667	952	5,619
Majority nonbank parents/ nonbank affiliates[d]	1,962[e]		(1,935) Total		246	4,143	902	5,045

Notes:
[a]Extrapolated from the nonbank figure of 952 by multiplying by 4857/4667. This may still involve a slight underestimate because U.S. licensors with no foreign affiliates are not included in this statistical universe. Also see Table 9.4 for more detail.
[b]Pages 95 and 103 of the Survey.
[c]Page 217. A breakdown of Service Charges and Rentals is unavailable in this category.
[d]Page 426. A breakdown of Service Charges and Rentals is unavailable in this category.
[e]See also page 372 of the Survey.

Source: *U.S. Direct Investment Abroad, 1977*, pp. 95, 103, 217, 372, 426.

TABLE 9.2
Relationship between Gross Receipts and Receipts Netted Against Affiliates, 1977
($ millions)

	Direct Investment Fees and Royalties		
	Gross Receipts	*Payments to Affiliates*	*Net Receipts from Affiliates*
All parents[a]	4,857	974	3,883
Nonbank parents/ nonbank affiliates[b]	4,667	971	3,697
Majority nonbank parents/ nonbank affiliates[c]	4,143	863	3,280

Notes;
[a]Page 95 of Survey.
[b]Page 217 of Survey.
[c]Page 426 of Survey.

Source: *U.S. Direct Investment Abroad, 1977.*

by 1.25 to estimate gross receipts from affiliates in Table 9.5 for the years 1972-81. This of course cannot produce a highly accurate estimate because the ratio of gross to net, 1.2508 in the benchmark survey for 1977, must vary slightly from year to year.* But the objective here is to give the reader an overall idea of the magnitude of licensing receipts, to compare these with equity-related receipts, and to show what portions of the total licensing receipts can be described as coming from independent or semi-independent foreign licensees.

The essential point in Table 9.5 is the little-realized fact that gross receipts from licensing were well over $8 billion in 1981. How does this compare with the returns that U.S. firms were extracting by way of dividends and repatriated earnings of unincorporated affiliates? This is shown in Table 9.6. Here again, for want of data, the repatriated earnings of unincorporated affiliates are estimated

*The ratio for other than benchmark years is not available without "costly computer programming," as stated in correspondence with the Bureau of Economic Analysis (BEA) of the U.S. Department of Commerce.

TABLE 9.3
Royalties and License Fees Netted against Affiliates[a]
($ millions)

Year	*Affiliated*	*Unaffiliated*
1981	3,650	1,282[b]
1980	3,693	1,170
1979	3,002	1,068
1978	2,697	1,055
1977	2,173[c]	923
1976	1,956	844
1975	1,886	757
1974	1,649	751
1973	1,376	712
1972	1,065	655

Notes:
[a]The netting of course does not apply to unaffiliated licensees.
[b]Provisional estimate.
[c]1977 figures are from the Benchmark Survey.

Sources: *U.S. Direct Investment Abroad, 1977* for 1977 data. Issues of *Survey of Current Business*, for other years. Unaffiliated data from Balance of Payments Division of the BEA by correspondence. *Selected Data on U.S. Direct Investment Abroad, 1966-78*, and its update.

by assuming that unincorporated affiliates repatriated in any year the same fraction of earnings that incorporated affiliates declared as dividends that year.* This data collection problem is described by the Commerce Department as follows: "In practice, unincorporated affiliates . . . have difficulty separating the portion of their total earnings that is remitted from the portion that is not."[12] Just as dividends comprise only the repatriated fraction of foreign earnings of incorporated affiliates, so all the earnings of unincorporated affiliates are not repatriated, regardless of their treatment in the data issued. "All earnings of unincorporated affiliates are treated as if they are remitted by U.S. parents even if in fact the earnings

*Informal conversation with Ralph Kozlow of the BEA indicated that this may be an unwarranted assumption. His opinion was that much, perhaps most, of the earnings of unincorporated affiliates may be repatriated.

remain abroad."[13] At any rate, my rough estimate leads to the conclusion that gross fees and royalties are of the order of 50 to 60 percent of the total repatriation on foreign direct investment equity holdings, as shown by the ratio B/A in Table 9.6.

While this is true in a national balance of payments sense, one cannot leap from there to the conclusion that in corporate strategy licensing is 60 percent as important as equity investment. Hardly so. After all, most licensing income is derived from majority controlled affiliates. The effective multinational tax treatment on licensing income and dividend type income is not identical. Political and in some cases even exchange inconvertibility risk is lowered by having royalties as well as dividends as alternative channels. For these reasons I would suspect that there is a bias toward signing licensing agreements. On the other hand, from Figure 9.3, we saw that a number of U.S. firms nevertheless simply do not bother to sign

TABLE 9.4
Fees and Royalties Netted against Affiliates[a]
($ millions)

Year	*Affiliated*	*Unaffiliated[b]*
1981	5,867	1,369[c]
1980	5,780	1,256
1979	4,980	1,147
1978	4,705	1,133
1977	3,863[d]	991
1976	3,530	906
1975	3,543	813
1974	3,070	806
1973	2,513	764
1972	2,115	703

Notes:
[a]Netting is of course only against affiliates.
[b]Estimated by multiplying Royalty & License Fees figures by 991/923.
[c]Provisional estimate.
[d]Page 95 of the Benchmark Survey.

Sources: Selected data in *U.S. Direct Investment Abroad, 1966-78* and updates. For 1977, the figures in *U.S. Direct Investment Abroad, 1977*. Balance of Payments Division of the BEA for unaffiliated figures. Issues of *Survey of Current Business*.

TABLE 9.5
Gross Receipts of Fees and Royalties, 1972-81

Year	*From affiliates*[a]	*Nonaffiliates*[c]	*Total*
1981	7,339	1,369[d]	8,708
1980	7,230	1,256	8,486
1979	6,229	1,147	7,376
1978	5,885	1,133	7,018
1977	4,857[b]	991	5,848
1976	4,415	906	5,321
1975	4,432	813	5,245
1974	3,840	806	4,646
1973	3,143	764	3,907
1972	2,646	703	3,349
Affiliates' receipts from unaffiliated foreigners (1977)			295
Grand Total, 1977			6,043

Notes:
[a]Estimated by multiplying netted receipts by 1.2508 = 4857/3883 from Benchmark Survey.
[b]Benchmark Survey, p. 95.
[c]Estimated by multiplying actual Royalties and Fees by 991/923.
[d]Provisional estimate.

Sources: Various issues of the *Survey of Current Business*. *U.S. Direct Investment Abroad, 1977*. Balance of Payments Division of the BEA.

formal agreements with affiliates, remaining content to extract returns from only the dividend channel (and of course from any eventual liquidation or sale value when the business is terminated).

In theory also it is difficult to disentangle the two types of returns. One can venture to say, with trepidation, that licensing income from an affiliate is a return on the firm's technical assets and past research, while dividends are a return on entrepreneurial capital. But this is an arbitrary distinction that may entertain tax lawyers; it does not yield much of prescriptive value to the corporate strategist. (We can say one thing with assurance: Licensing income is to be discounted to present value at a lower rate than dividend income.)

TABLE 9.6
A Comparison between Licensing Returns to the United States and Repatriated Earnings on Equity in Direct Investment

	1981	*1980*	*1979*	*1978*	*1977*	
Gross dividends[a]	10,541	11,783	10,042	8,762	8,217	
Estimated repatriation of earnings of unincorporated affiliates[b]	4,150	3,544	3,272	3,234	3,072	
Total return on equity position	14,691	15,327	13,314	11,996	11,289	(A)
Gross receipts of fees and royalties	8,708[c]	8,486	7,376	7,018	5,848	(B)
Ratio of B/A	0.59	0.55	0.55	0.58	0.52	

Notes:
[a]On common and preferred stock of incorporated affiliates before taxes.
[b]This is a rough estimate derived for want of better data by multiplying each year's figure for the earnings of unincorporated affiliates by the ratio of gross dividends over earnings of incorporated affiliates for that year.
[c]Provisional.

Sources: *Survey of Current Business*, various years, August issues.

In the case of a minority joint venture, a consideration emerges that is irrelevant with majority affiliates. Quite apart from the tax and risk minimization motivation for signing licensing agreements, there is now the simple fact that, while the U.S. firm will keep only a minority share of declared dividends, it will keep all of the negotiated licensing royalties and fees. The lower the percentage of equity held, the less control the U.S. firm has on the timing and size of the dividends – besides of course having less say on revenues, costs, and operating strategy of the business in general. It is clear that in all joint-venture situations, but particularly in minority joint ventures, there is every incentive to negotiate as large and lucrative a licensing payment as possible, as one would when negotiating with an independent licensee. This a priori reasoning leads to the conclusion that licensing income ought to be relatively more important in minority than majority affiliates. The data, however, show this is not so.

Table 9.7 shows the latest available information on the division of licensing income between majority and minority affiliates and unaffiliated licensees. The share contributed by minority affiliates is only 12 percent (or 15 percent of the affiliate total) whereas, according to Figure 9.3, minority affiliates had a 28 percent share in assets and 25 percent in employment. There are two plausible

TABLE 9.7
Approximate Division of 1977 Gross Receipts of Fees and Royalties (\$ millions)

				%
From majority (nonbank) affiliate		4,143[a]		71
From minority affiliates	714		12	
From unaffiliated foreigners	991		17	
	1,705			
		1,705		29
Total		5,848		100

Add affiliates' receipts from unaffiliated foreigners 295[b]

Notes:
[a]Page 426, *U.S. Direct Investment Abroad, 1977.*
[b]Page 372, *U.S. Direct Investment Abroad, 1977.*

Source: *U.S. Direct Investment Abroad*, 1977.

explanatory hypotheses. Either U.S. firms were not attempting sufficiently to sign up minority affiliates, or their joint-venture partners, aided in some countries by government policies such as in developing countries had sufficient negotiating strength to refuse licensing in addition to equity participation by the U.S. firm.* The point nevertheless remains that 29 percent of total licensing income was derived in 1977 from minority affiliates and unaffiliated parties. Assuming the same percentage has held constant since the last benchmark survey in 1977, that would amount to some $2.5 billion in licensing fees paid by such licensees in 1981 out of the gross total of $8.7 billion from all foreign licensees, a not inconsiderable sum in its own right or when compared to annual returns on foreign direct investment. At the corporate level such a comparison is, of course, dangerous for various reasons. A company's perception of risk associated with the two strategies is not identical. Dividends are, after all, only a portion of affiliate earnings, the rest being reinvested abroad for growth. And so on. However, the above analysis has demonstrated that in a balance of payments sense, the role of licensing income in general and of arm's-length licensing in particular is larger than commonly assumed.

THE STRATEGY CHOICE AS A FUNCTION OF COUNTRY AND INDUSTRY VARIABLES

The above discussion placed in perspective the relative importance of licensing and foreign equity returns. But how does the choice, as a strategic decision, vary as a function of foreign country and U.S. industry characteristics? This is the subject of the following analysis. At the outset it should be noted that the data for licensing from here on will refer *only* to unaffiliated licensing income (that is, where the U.S. firm has less than 10 percent equity stake in the licensee). Data for licensing with minority joint ventures, where

---

*The share of developing countries in total fees and royalties in 1977 is 14.1 percent in minority affiliates and 15.8 percent in all affiliates, much lower percentages than their share in dividends. It is significant to note, however, that in 1981 the developing country share in (netted) fees and royalties was up to 21 percent. This would tend to support the former rather than latter hypothesis, because no one will deny that the bargaining power of developing countries has increased in the last five years.

the equity stake is between 10 and 50 percent, are simply unavailable in the disaggregation needed for this analysis. Nevertheless the unaffiliated licensing data will comprise an adequate surrogate for licensing used as a primary strategy in its own right, as an alternative to foreign direct investment. (Moreover, the entire issue of a tax bias in licensing fee determination is thus avoided.) The data are for three years, 1977-80, to check the robustness of the cross-sectional analysis over time. Table 9.8 shows the hypothesized

TABLE 9.8
Variable Definitions

Symbol	*Description*	
Dependent Variables		
σ	Ratio of U.S. receipts of royalties and licensing fees from unaffiliated firms *over* various measures of direct investment activity, described below:	
	For Country i or for U.S. Industry j and World Region k	
σ_i^1 or σ_{jk}^1	Ratio of U.S. receipts of unaffiliated royalties and licensing fees *over* U.S. "direct investment position"	
σ_i^2 or σ_{jk}^2	Ratio of U.S. receipts of unaffiliated royalties and licensing fees *over* U.S. direct investment "income"	
σ_i^3 or σ_{jk}^3	Ratio of U.S. receipts of unaffiliated royalties and licensing fees *over* the sum of U.S. direct investment "income" and affiliate licensing income	
$UNRO_i$ or $UNRO_{jk}$	U.S. receipts of unaffiliated royalties and licensing fees ($'000s)	
Independent Variables		*Expected Sign*
$PERS_i$	Research personnel per 100,000 population in country i	+
$RANDEX_i$	R & D expenditures in country i, $	+
PERCAP	Per capita GDP in country i or region k, as indicated	–
$TYPE_i$	Dummy variable indicating extent of country i government intervention or restriction on foreign investment	+

continued

TABLE 9.8, Continued

Symbol	*Description*	*Expected Sign*
Independent Variables, continued		
	(8 = maximum and 1 = minimum) from Gottfried and Hoby (1978)	
$RISK_i$	Investment environment score (0 to 100 best) from *Business International* (1981)	–
$CONC_i$ or $CONC_{jk}$	Fraction of U.S. companies that have received specific incentives on investment, from U.S. Dept. of Commerce (1981)	–
$MANUF_i$	Percentage of manufacturing in GDP (0 to 100)	–
PAT_i	Patents in force, thousands	+
DIP_i or DIP_{jk}	Direct investment position, $ billion	–
NET_{jk}	Ratio of net income to sales in direct investment affiliates	–
RAN_j	R & D expenditures per firm in industry j	–
MEM_j	Ratio of managerial workers in total employment in U.S. industry j	Nonzero
REM_j	R & D scientists and engineers per firm – average in U.S. for industry j	Nonzero
$MEMP_j$	Number of managerial employees in U.S., thousands, industry j	Nonzero
$REMP_j$	R & D scientists and engineers, thousands, industry j	Nonzero
$ASSETS_j$	Assets per firm of firms investing abroad, in industry j, $ billion [– in Eqn (3); + in Eqn (4)]	–; +
$RAND_j$	R & D expenditures by U.S. parents of equity affiliates, industry j, $ billion	–

relationship between the relative use of licensing and independent variables describing certain industry and country characteristics.

As we saw, a drawback to licensing in some nations is the fear of licensee competition in third countries, or even in the licensor's home nation, as illustrated by the experience of some U.S. companies who licensed in Japan. In general, though, with patents a licensee can be perfectly legally restricted to a particular territory. This would suggest that, ceteris paribus, patented technologies are more amenable to licensing. This is to be tested. (The opposite argument that patents indicate "higher" technology, which a firm would prefer to share only with affiliates, is weaker. Stoneman actually found a negative correlation between the number of patents and the R & D cost per patent.[14] Other studies show little correlation between the value of a technology and its being patented or not.)

I hypothesize next that, the greater are R & D expenditures leading to forms of technology more complex and proprietary, the more the firm has to lose and, to that extent, will avoid licensing, as Wilson or Davidson and McFetridge indicate.[15]

Besides the monopolistic advantage derived from the size of a company's R & D budget or its R & D intensity, there may also be important expertise in administration and marketing. Several studies, for example, those by Dunning and Kobrin find foreign direct investment positively correlated with measures such as the share of managers in total employment, or the years of education of employees.[16] Since these studies did not address licensing, we have no a priori information on how such measures would correlate with the ratio of licensing to direct investment, the principal focus of this study. One could even argue, for instance, that the systematic organization of information in a company would favor its ability to license. Let us make no a priori judgment, except to propose a non-null hypothesis.

Finally, on the question of the relationship between firm size and propensity to license versus invest, papers by Casson, and Buckley and Davies suggest the hypothesis that smaller firms, lacking the managerial and financial resources to make direct investment, would be more prone to favor licensing.[17]

As for country variables, in nations where direct investment earnings attract a higher tax rate or are eligible for fewer incentives, one can hypothesize licensing to increase relative to equity investment, ceteris paribus. If onc cxamines the effects of country risk on the strategy choice, in licensing there are fewer or no assets at stake and royalties being pegged to turnover; they are more stable than dividends, the latter being declared only out of profits, if any. Many agreements include minimum royalty clauses and front-end payments. Moreover, the planning horizon is nearer and net cash inflows begin sooner than in the case of equity investment. Last, in several countries there are fewer restraints on the convertibility of contractual licensing fees than on dividends. These reasons lead to the hypothesis that as country risk increases, licensing would tend to be favored over investment, ceteris paribus, in a cross-sectional sense.

As a country's indigenous technical capability increases, as measured by indicators such as its R & D expenditures or research personnel per capita, other things being equal, there would be a relatively greater use of licensing, I hypothesize. Appropriable profit available to the foreign investor would be lower as local companies and technologies compete more effectively and hold their own against foreign investment. By the same token, local companies as licensees would be in a better position to absorb the foreign technology, with a lower cost to the licensor to transfer the technology and, at the same time, be in a position to offer a higher compensation to the licensor. (Both these assertions are verified in an empirical study by Contractor.)[18] The licensee's ability to compete in third countries as local technical capabilities are enhanced would appear to counteract the above hypothesis. However, studies show that in practice the cases where a licensee does actually export or produce in a third country in competition with the licensor are infrequent, except recently, in a few nations like Japan.

How does the level of economic development, as measured by indicators such as per capita gross domestic product (GDP), affect foreign investment and licensing? Several studies (for example, Davidson, Kobrin, and Green and Cunningham,[19] show a positive correlation between U.S. manufacturing direct foreign investment and per capita GDP. (Licensing has not been analyzed.) But what does this measure? Caution is required. If per capita GDP measures market potential or size for manufacturing technologies and products (this study's focus), we can connect with the analysis in Casson or Buckley and Davies[20] that licensing would tend to be favored over investment in less developed markets, and vice versa, other things being equal. It is argued that a less developed market will provide a smaller total profit on an equity investment; but the costs of a capital investment in financial and managerial terms may not be very much less than in a large market if there are minimum scale requirements or indivisibilities. A smaller market also favors licensing because it represents a lower opportunity cost to the licensor (in abdicating the market to the licensee). On the other hand, to the extent that per capita GDP measures, through colinearity, other factors such as political risk or investment climate, there are alternative underlying explanations, albeit with the same prediction as to the corporate choice of methods. For instance, ceteris paribus, licensing is said to be preferred over investment, the greater the perceived country risk. But if risk is negatively correlated to per capita GDP, then the hypothesis of increased propensity to use licensing in developing nations could be explained as much on the basis of risk and investment climate as on market potential. Since the predicted direction of the statistical relationship is the same in either explanation, the alternative theoretical rationales are, at any rate, convergent; but there could be an operational difficulty with possible multicolinearity, to judge from the experience of other cross-sectional studies such as that of Root and Ahmed.[21]

Last, I will examine the relationship between patent filings in a country and the corporate strategy choice. I argue here that the greater the patent intensity, the greater the relative propensity to use licensing. As usual, one has to be circumspect in statistical analysis about the direction of causality, if any. Does greater patent intensity lead to greater use of licensing, or vice versa? One line of reasoning supporting the former is that higher prices are paid for patented technologies, ceteris paribus, as confirmed in Contractor,[22] and the patent system legally enables market segmentation, obviating the fear, real or imagined, of licensee competition in third countries. At the same time, through the licensee's monopoly, a high rent or profit is available to the licensee and, therefore, to the licensor. It will be enough in this exploratory empirical study to establish at least a significant relationship between the variables.

MODEL SPECIFICATIONS

Two basic models are tested; the first using only country variables, the second with both U.S. industry and foreign country variables. Table 9.8 defines each variable. Because of limitations of

Commerce Department data, such as suppression of entries when corporate confidentiality might be violated, the number of usable variables and cases is restricted. For instance, on a U.S.-industry/country matrix, complete data are not available at that level of disaggregation. A complete data set can be assembled for 30 countries taken alone. (Please refer to the appendix for a list of the countries, data sources, and limitations.) With an industry classification, however, (food products, chemicals, metals, nonelectrical machinery, electrical machinery, transport equipment, other manufacturing, and petroleum), only three countries can be broken out (Canada, the United Kingdom, and Japan); whereas, complete data for the rest are available only as regions (EEC-6, Latin America, and Australia, New Zealand, and South Africa, as a group).

Country Variables: (for country i)

$$\sigma_i = f\,(PERS_i, RANDEX_i, PERCAP_i, TYPE_i, RISK_i, CONC_i, MANUF_i, PAT_i) \quad (1)$$

$$UNRO_i = f\,(CONC_i, RANDEX_i, TYPE_i, PAT_i, DIP_i) \quad (2)$$

Industry/Region Variables: (industry$_j$ and Region$_k$)

$$\sigma_{jk} = f\,(RAN_j \textit{ or } REM_j, NET_{jk}, CONC_{jk}, MEM_j, PERCAP_k, ASSETS_j) \quad (3)$$

$$UNRO_{jk} = f\,(REMP_j \textit{ or } RAND_j, MEMP_j, CONC_{jk}, ASSETS_j, DIP_{jk}) \quad (4)$$

In equations (1) and (3), the main focus of the analysis, the dependent variable is the *ratio* of income from unaffiliated licensees over three distinct measures of direct investment activity. These are: 1) U.S. direct investment position, as defined by the Commerce Department; 2) U.S. direct investment "income"; and 3) direct investment "income" plus royalties and fees paid by affiliates. The reader is referred to any August issue of *Survey of Current Business* for details. Being a stock measure, the direct investment position is a less volatile index over time and is to be preferred somewhat to the "income" measures. Data for three years are analyzed to check for stability of results over time. (Please refer to the Appendix for further details.)

As an additional exercise, instead of the above ratios, the absolute value of dollar licensing fees and royalties from unaffiliated licensees was used as a dependent variable in equations (2) and

(4) and regressed against certain absolute country and industry measures.

Using the country data, a priori, I propose that σ_i is positively associated with indigenous technological capabilities in foreign countries (expressed by the variables PERS and RANDEX) with the degree of government intervention or controls on investment (expressed by the variable TYPE) which Contractor, Dunning,[23] and others claim favor the licensing mode. It is expected that σ_i will be negatively associated with the level of economic development and industrialization (expressed with caveats, discussed above, by PERCAP and MANUF), and negatively associated with the quality of the investment environment (expressed by RISK) and concessions given to foreign direct investment (measured by CONC).

In Equation (2), I propose that the absolute value of unaffiliated licensing income, UNRO, is positively correlated, ceteris paribus with R & D expenditures, government intervention or restriction on foreign investment, and patent filings and negatively related to the level of concessions to equity investment and U.S. direct investment position in the country (DIP). This last relationship presumes that arm's-length licensing and direct investment behave as substitutes. If, on the other hand, we actually find a positive relationship, then unaffiliated licensing would have to be viewed as complementary to equity investment, perhaps from a foreign market "experience effect," proposed by Davidson.[24]

Theory does not tell us of the functional form of the relationship between the dependent and independent variables. An OLS form of regression is used, quite satisfactorily as it turns out. (A single and double logarithmic form gives essentially similar results.)

In Equation (3), I propose that σ_{jk} is negatively associated, ceteris paribus, with CONC and PERCAP, as above, and negatively associated with the profitability of direct investment in the region, as measured by the variable NET, once again presuming that independent licensing and direct investment behave as substitutes. For REM and MEM, as alternate indicators of internal firm advantages, let us posit simply a non-null relationship since the a priori theoretical arguments are weak on this score. A negative sign for ASSETS is proposed from the hypothesis that larger international firms are better able to internalize technology transfers and are likely to use less licensing relative to investment.

In Equation (4), for absolute unaffiliated licensing income, the sign for DIP is again hypothesized to be negative; but the sign for ASSETS is expected to be positive, on the assumption that *absolute* level of international licensing will be higher the larger the average firm size in the industry. As indicated, it is regrettable that such industry averages have to be used. Further disaggregation of nonaffiliate licensing data is simply unavailable (see appendix).

RESULTS

For cross-sectional data, the results in general are gratifying, in that the percent variation explained (R^2) is high in most equations, overall equation F-values are strong, the coefficients show stability over different years, and the signs of the coefficients, in most cases, are as hypothesized.

Results from Equation (1) are shown in Table 9.9. At the outset, it should be noticed that the variable RISK turned out to be strongly colinear with PERCAP and had to be dropped. RISK is a composite index, based on numerous underlying factors, which ranks countries in terms of their desirability as equity investment locations.[25] If, however, such measures turn out to be colinear with per capita GDP, then one cannot but wonder whether corporations that buy this data are spending their money well. A too hasty indictment must be tempered, however, with the observation that the colinearity was observed only in this sample and may not occur in general. This remains an open question.

Data from PAT were available for only 19 of the 30 nations, and thus the degree of freedom was reduced. The 11 countries dropped include 6 developing and 5 developed countries (see Appendix), so that at least superficially, there seems to be no bias on that score. In estimates 2, 4, 6, and 7, PAT is dropped in order to use data for all 30 countries.

The results confirm the hypothesis that the relative propensity to use licensing increases with the technology-receiving country's indigenous technical capabilities, as measured by research expenditures, and research personnel normalized by population. Again, as hypothesized, the relative proportion of licensing decreases with the level of economic development, measured by per capita GDP,

TABLE 9.9
Regression Statistics Equation (1) for σ_i (for years 1977 and 1980 only)[a]

No.	Year	Dependent Variable	Const.	PERS	RANDEX	PERCAP	TYPE	CONC	MANUF	PAT[b]	d.f.	F Value	R^2
1	1977	σ_i^1	0.40	0.85	0.27	–0.49	–0.58	–0.16	–0.85	0.15	9,7	3.77*	0.75
				(2.53)**	(1.70)†	(2.15)*	(0.29)	(0.71)	(1.74)†	(0.54)			
2	1977	σ_i^1	0.32	0.54	0.30	–0.18	0.10	0.57	–0.19		6,22	4.43**	0.55
				(2.15)**	(2.79)**	(1.36)†	(0.98)	(0.58)	(0.68)				
3	1980	σ_i^1	0.50	0.77	0.26	–0.54	–0.95	–0.26	–0.85	0.10	7,9	3.75*	0.74
				(2.32)**	(1.66)†	(2.42)**	(0.48)	(1.16)	(1.79)†	(0.36)			
4	1977	σ_i^2	0.99	0.82	0.66	–0.21	0.46	d	–0.26		5,24	1.11	0.19
				(1.70)*	(0.35)	(0.89)	(0.26)		(0.54)				
5	1980	σ_i^2	0.35	0.49	0.24	–0.34	0.44	–0.15	–0.73	0.37	7,9	3.01†	0.70
				(1.83)*	(1.87)*	(1.90)*	0.78)	(0.83)	(1.89)*	(0.16)			
6	1980	σ_i^2	0.51	0.29	0.22	–0.13	0.58	d	–0.17		5,22	3.95**	0.47
				(1.49)†	(2.65)**	(1.36)†	(0.84)		(0.80)				
7	1977	σ_i^3	Overall equation not significant using all variables[c]										

8	1980	σ_i^3	0.25	0.32 (1.71)†	0.17 (1.94)*	−0.24 (1.91)*	−0.19 (0.17)	−0.11 (0.79)	−0.56 (2.07)*	0.11 (0.07)	7,9	3.06†	0.70

Notes:
[a]Results for 1978 are roughly similar.
[b]Data for PAT available for only 19 of the countries. When PAT is *not* used, number of cases = 30.
[c]However signs of coefficients congruent with hypotheses except for CONC.
[d]Variable too insignificant to be included.
Significance levels:
**0.025
*0.05
†0.10
for one- or two-tailed tests as specified. t-values in parentheses.

and with the level of industrialization, measured by the percentage of manufacturing in GDP, although in the latter case, the statistics are weaker. For PAT, the sign is consistently positive, as predicted, indicating that patent filings are positively associated with relative use of licensing. The direction of causality in this instance remains an open question, however. For CONC and TYPE, the signs are as predicted only in a majority of estimates; nor are the t-values significant. There is insufficient evidence to reject the null hypothesis for these two variables.

The absolute level of U.S. licensing activity, as measured by unaffiliated licensing income UNRO, is confirmed to be positively associated with R & D expenditures in the nation at better than the 0.025 significance level, and with patent filings. The negative sign of DIP would tend to confirm the idea that nonaffiliate licensing is a strategy substitute for, rather than a complement to, direct investment. The sign of TYPE is positive, as expected, supporting weakly the idea that increased host government intervention and regulation of direct foreign investment tends to favor the alternative market entry strategy of licensing. The sign of CONC is consistently negative, supporting the idea that concessions given to equity investment lower licensing, as predicted. Overall F-values and R^2 of the equations are all high.

As far as the industry/region data are concerned, it should be noted at the outset that, in estimating Equation (3) in Table 9.11, RAN (R & D expenditures per firm) was dropped, being colinear with REM (R & D employees per firm). In any event, REM is not significant, so that we are unable to say that R & D intensity does or does not influence the corporate choice of strategy. The ratio of licensing over direct investment is shown to be positively associated, ceteris paribus, with the ratio of managerial employees in total employment MEM (at the 0.025 or 0.05 levels), and to be negatively associated with the incidence of concessions granted to direct equity investment (at the 0.025 level), as expected. The sign for PERCAP is consistently negative, as anticipated; but the coefficients are not significant. The sign for ASSETS is opposite to the hypothesis that the relative use of licensing compared to investment would increase in smaller firms, ceteris paribus. The variable was not significant, however, in any estimate. The paradoxical result is for the variable NET. I hypothesized that, assuming that direct investment and licensing behave as substitutes, ceteris paribus,

TABLE 9.10
Regression Statistics: Equation (2) for $UNRO_i$

No.	*Year*	*Dependent Variable*	*Const.*	*RANDEX*	*TYPE*	*CONC*	*DIP*	*PAT*[a]	*d.f.*	*F Value*	R^2
9	1977	$UNRO_i$	−10885.60	18.43	2259.18	−4565.65	−2.35	271.67	5,13	14.04**	0.84
				(4.59)**	(0.63)	(0.11)	(1.54)†	(2.09)**			
10	1977	$UNRO_i$	−8032.59	23.79	2325.25	−4935.26	0.86		4,25	23.03**	0.79
				(8.78)**	(1.03)	(0.19)	(0.10)				
11	1978	$UNRO_i$	−18760.21	24.37	3524.46	−3722.77	−2.49	291.25	5,13	12.62**	0.83
				(4.75)**	(0.76)	(0.07)	(1.36)†	(1.76)*			
12	1978	$UNRO_i$	−8478.95	29.56	2936.44	−10989.15	−0.14		4,25	22.66**	0.78
				(8.66)**	(1.04)	(0.35)	(0.14)				
13	1980	$UNRO_i$	−22635.99	23.97	4562.44	364.86	−2.16	315.08	5,12	9.83**	0.79
				(4.06)**	(0.87)	(0.00)	(1.34)†	(1.67)†			
14	1980	$UNRO_i$	−10789.27	29.98	3381.30	−7294.48	−0.17		4,25	17.94**	0.74
				(7.79)**	(1.06)	(0.20)	(0.20)				

Notes:
[a]Data for PAT available for only 19 of the countries. When PAT is not used, the number of cases is 30.
Significance levels:
**0.025, *0.05, †0.10 — for one- or two-tailed tests as specified. t-values in parentheses.

TABLE 9.11
Regression Statistics Equation (3) for σ_{jk}

No.	Year	Dependent Variable	Const.	REM	NET	CONC	MEM	PERCAP	ASSETS	d.f.	F Value	R^2
15	1978	σ^1_{jk}	–0.42	–0.63	2.51	–0.30	4.39	–0.16	0.78	6,35	13.63**	0.70
				(0.08)	(6.86)**	(2.32)**	(2.40)**	(0.19)	(1.20)			
16	1978	σ^2_{jk}	–1.26	–0.16	7.63	–0.16	18.42	–0.15	0.20	6,34	2.98*	0.34
				(0.37)	(1.79)†	(2.54)**	(2.02)*	(0.37)	(0.61)			
17	1978	σ^3_{jk}	–1.01	–0.70	5.68	–0.10	13.51	–0.17	0.85	6,28	2.75*	0.37
				(0.15)	(1.76)†	(2.24)**	(1.92)*	(0.05)	(0.34)			
18	1979	σ^1_{jk}	–0.27	–0.70	0.41	–0.28	4.10	–0.37	0.48	6,35	2.43*	0.29
				(0.83)	(1.13)	(2.16)**	(2.28)**	(0.45)	(0.75)			
19	1979	σ^2_{jk}	–7.90	2.45	76.18	–0.13	36.83	0.82	1.64	6,34	42.25**	0.88
				(1.62)	(12.60)**	(0.64)	(1.24)	(0.61)	(1.57)			
20	1979	σ^3_{jk}	–1.62	0.12	11.84	–0.11	16.71	–0.14	0.22	6,29	15.97**	0.77
				(0.31)	(7.54)**	(2.06)**	(2.14)**	(0.38)	(0.75)			

Note:
Significance levels:
**0.025
*0.05
†0.10
for one- or two-tailed tests as specified. t-values in parentheses.

one might reasonably expect that the ratio of net income to sales in direct investment affiliates NET would be negatively associated with the dependent variable; that is, the more profitable equity investment is, the lower one might expect the ratio of licensing over equity investment to be. The opposite was found. Perhaps the data are simply anomalous, peculiar to the time of this cross-sectional study.

Table 9.12 shows the results of estimating Equation (4). At the outset, REMP, RAND, and MEMP were all found to be colinear, and two of the variables were dropped as a result. While the signs of the relationships are as expected, only CONC and REMP are significant (at the 0.025 level). Again, however, the sign of ASSETS is contrary to a priori expectation, but the variable is not significant. The purpose of Table 9.13 is to show essentially similar results when the data are partitioned into 2 roughly equal halves, 16 industrial nations and 14 developing countries. The model remains robust in the industrial nation subsample, but the statistics are very weak in the developing country subsample.

CONCLUSIONS AND POLICY IMPLICATIONS

The results lend credence to the idea that the strategy choice of arm's-length licensing versus direct investment made by U.S. multinational firms is influenced (and a large portion of statistical variation explained) by both country and industry characteristics, but in a complex way. For instance, the ratio of licensing to investment (in the cross-sectional analysis) increases with technical capability in a country, a variable that can indeed be influenced in the long run by government policy, as in Japan. Licensing, however, decreases in relative importance with higher per capita GDP and industrialization, ceteris paribus. Similarly, we found weaker support for the idea that the proportion of licensing increases, ceteris paribus, with government scrutiny and regulation of direct investment and decreases, on the other hand, as more concessions and incentives are offered on direct investment.

This was exemplified by two countries that sometimes were outliers, Japan and the United Kingdom. The ratio σ for Japan was well above its predicted value in many estimates and that for Britain under. A similar situation was found in empirical studies by Kobrin

TABLE 9.12
Regression Statistics Equation (4) for $UNRO_{jk}$

No.	*Year*	*Dependent Variable*	*CONST.*	*REMP*	*CONC*	*ASSETS*	*DIP*	*d.f.*	*F Value*	R^2
21	1978	$UNRO_{jk}$	53.77	0.40	–1.05	–15.14	–0.11	4,43	12.69**	0.54
				(4.79)**	(2.99)**	(1.25)	(0.52)			
22	1979	$UNRO_{jk}$	52.76	0.40	–0.97	–19.10	–0.95	4,43	12.23**	0.53
				(2.76)**	(4.26)**	(1.57)	(0.44)			

Note:
Significance levels: **0.025, for one- or two-tailed tests as specified. t-values in parentheses.

TABLE 9.13
Regression Statistics of Some Tests on a Subsample of Sixteen Industrial Nations[a]

No.	Year	Dependent Variable	Const.	PERS.	RANDEX	PERCAP	TYPE	MANUF	d.f.	F Value	R^2
23	1977	σ_i^1	0.24	0.45	0.37	–0.29	0.95	–0.68	5,10	3.68*	0.65
				(1.38)†	(2.30)**	(1.57)†	(0.73)	(1.10)			
24	1980	σ_i^1	0.26	0.40	0.38	–0.29	0.83	–0.77	5,10	3.98*	0.67
				(1.33)	(2.53)**	(1.66)†	(0.68)	(1.33)			
25	1980	σ_i^2	0.22	0.22	0.30	–0.19	0.45	–0.65	5,10	3.40*	0.63
				(0.95)	(2.63)**	(1.48)†	(0.49)	(1.48)†			

Notes:
[a]The above three equations exemplify the similarity of results for the subsample with the overall sample. Results for 1978 and other versions of σ are also similar.

Significance levels:
**0.025
*0.05
†0.10
for one- or two-tailed tests as specified. t-values are in parentheses.

and Contractor.[26] U.S. international firms clearly favor the direct investment route in Britain, where, to go back to the data, there are far more concessions and incentives and a less stringent regulatory climate for foreign investment than in Japan. On the other hand, on indicators of indigenous technical capability, Japan is significantly higher. Clearly, governmental policy influences the choice of business methods.

Licensing was shown to be positively associated with the number of patents filed in a country, but the direction of causality is unknown. A plausible explanation may be that patents increase revenues extractable from licensing, and 1) obviate an important drawback of licensing, the fear of licensee competition in third countries, or 2) increase global licensing revenues by effective market segmentation with the international firm as a discriminating monopolist. If we accept this hypothesis, then a weakening of the patent system sought by some developing countries would only diminish international licensing and drive firms to greater trade secrecy and "internalization" of the technology in controlled equity affiliates, as Casson posits.[27] This may not be advantageous to technology-receiving countries because studies, as well as regulatory behavior in Japan, Mexico, and India, among other countries that favor licensing over investment, suggest a belief that licensing may involve a lower long-run national cost of acquiring technologies as compared to equity investment, at least for some nations.

As for corporate policy, the absolute level of licensing in a country was shown to be generally negatively related to the level of direct investment, from which one may conclude that the two are strategy substitutes rather than complements. There are weaker indications that the ratio of licensing to investment increases with the ratio of managerial employees in total employment but is inversely related to research intensity.

Many of the predictions of theory, for example, those by Casson and Magee, are borne out.[28] Firms may prefer direct investment under most conditions, but licensing, while remaining minor, may increase its role as the environment is more restrictive toward investment, as the technology is less research intensive, more codified, patented, and transferable, and as the recipient country has greater indigenous technical capabilities.

APPENDIX: DATA SOURCES AND LIMITATIONS

Data for the dependent variables were obtained from the BEA and from *U.S. Direct Investment Abroad, 1977* and *Selected Data on U.S. Direct Investment Abroad*, both publications of the BEA. Their definition of nonaffiliated licensing is that ownership or control by the U.S. licensor not exceed 10 percent of the licensee enterprise's voting securities. This omits an important segment of licensing, where the U.S. firm has between 10 and 50 percent of equity, on which data are not available. This and other factors, such as the exclusion from the figures of auxiliary business that often surrounds the core agreement for royalties and fees, mean that the BEA figures for nonaffiliated licensing considerably understate its strategic importance to U.S. companies. Nevertheless, they are adequate for a statistical analysis.

The confidentiality criteria used do not have the result of deleting too many numbers in the direct investment portion of the data (namely, direct investment position or direct investment "income"), even when the data are disaggregated to the country level or the BEA equivalent of a three-digit standard industrial classification. However, there is a problem in disaggregating the unaffiliated royalties and licensing fees. In computing the ratio σ, complete matched data were available for only a total of 30 nations (16 industrial advanced nations and 14 developing countries) – Argentina, Australia, Belgium, Brazil, Canada, Chile, Colombia, Denmark, Egypt, France, West Germany, India, Indonesia, Ireland, Italy, Japan, Mexico, the Netherlands, New Zealand, Nigeria, Norway, Panama, Peru, the Philippines, South Africa, Spain, Sweden, Switzerland, the United Kingdom, and Venezuela. Iran and Iraq, for which data were available, were, however, not included because oil revenues considerably distort not only the U.S. direct investment figures but also independent variables such as per capita GDP. Because suppression of data occurs most often in very small countries – with, for example, fewer than three U.S. companies reporting – it is reasonably accurate to say that the 30 nations above include most of the significant countries from which U.S. firms derive nonaffiliated licensing income.

The problem of disaggregation was even more acute at the industry level, and breakdown on an industry/region matrix was available only by BEA "major industry group." (For details, see

U.S. Direct Investment Abroad 1977, pp. 8-9, 494-516.) These were seven groups under "manufacturing," namely food and kindred products; chemicals and allied products; primary and fabricated metals; machinery, except electrical; electric and electronic equipment; transportation equipment; other manufacturing; and one nonmanufacturing category, petroleum. These eight groups cover the great bulk of investment and licensing activity. For instance, they comprise 87 percent of the total direct investment abroad in 1977 of U.S. industries other than banking, finance, and trade. On the industry/region matrix, confidentiality constraints on the licensing data again enable only a six-region disaggregation: Canada, the United Kingdom, the European Economic Community (original six countries), Japan, Latin America, and Australia, New Zealand, and South Africa as a group. Even so, there are some suppressed entries, and, instead of 48 cases (8 industries × 6 regions), the usable number of cases is closer to 40.

Sources of data on the independent variables are as follows:

Variable	*Source*
PERCAP, MANUF	*Worldwide Economic Indicators Series*, Business International Corporation, New York (1975 prices)
PERS, RANDEX	*UNESCO Yearbook*
TYPE	Gottfried and Hoby (see bibliography)
RISK	Business International (see bibliography)
CONC	*U.S. Direct Investment Abroad, 1977*, Table II-K1, page 168
PAT	World Intellectual Property Organization
REM, MEM, REMP MEMP, NET, DIP ASSETS	*U.S. Direct Investment Abroad, 1977* (Group II data for Non-bank affiliates of non-bank parents)

For REM, MEM, REMP, and MEMP, the data relate to U.S. investors abroad rather than to U.S. industry in general. This presupposes that the universe of U.S. arm's-length licensors is roughly identical to the universe of U.S. companies investing abroad (rather than comprising distinct subsets of U.S. industry in general). There are some indications that this assumption is accurate, although it remains to be conclusively proven empirically.

The 11 countries dropped when PAT is included are Chile, Indonesia, Nigeria, Panama, Peru, and Venezuela; and Australia, Italy, New Zealand, Spain, and Switzerland – a roughly equal number of developing and industrial countries.

Ordinary linear least squares models may be a problem if the dependent variable has both an upper and lower bound and if there is significant bunching of values. There is only a lower bound of zero in these equations, and there is no bunching of values. In any event, when the dependent variable is log-transformed, the signs of the coefficients conform with the Ordinary Least Squares (OLS) model. Since theory does not specify the form of the function, if OLS is satisfactory further computer trials to squeeze out an incremental R^2 gain may not be warranted. In any case I am using the regressions not so much for an econometric predictive ability as to identify the subset of relevant independent variables that influence the corporate choice in the anticipated direction.

NOTES

1. M. Sato and R. Bird, "International Aspects of the Taxation of Corporations and Scholars," *IMG Staff Papers* no. 22, July 1975.
2. G. Kopits, "Intra-Firm Royalties Crossing Frontiers and Transfer-Pricing Behavior," *Economic Journal*, December 1976, pp. 791-803.
3. M. Finnegan and R. McCarthy, "U.S. Tax Considerations in International Technology Transfers," *Licensing Law and Business Report*, April 1979, pp. 101-12; May 1979, pp. 113-19.
4. M. Casson, *Alternatives to the Multinational Enterprise* (New York: Holmes and Meier, 1979).
5. D. Teece, *The Multinational Corporation and the Resource Cost of International Technology Transfer* (Cambridge, Mass.: Ballinger, 1977); F. Contractor, *International Technology Licensing: Compensations, Costs and Negotiation* (Lexington, Mass.: D. C. Heath, 1981).
6. D. Teece, "The Market for Knowhow and the Efficient International Transfer of Technology," *Annals of the Academy of Political and Social Sciences* (November 1981).
7. Casson, *Alternatives to the Multinational Enterprise*.
8. R. Caves, "International Corporations: The Industrial Economics of Foreign Investment," *Economica* 38 (February, 1971): 1-27; S. Hymer, *The International Operations of National Firms: A Study of Direct Investment*, Ph.D. dissertation, Massachusetts Institute of Technology, 1960.

9. U.S. Department of Commerce, *U.S. Direct Investment Abroad, 1977* (Washington, D.C.: Government Printing Office, 1981), p. 3.

10. Ibid., pp. 190, 191, and 400.

11. For a further discussion of the conceptual separation and a cross-sectional statistical analysis based on this, see F. Contractor, *International Technology Licensing.*

12. U.S. Department of Commerce, *U.S. Direct Investment Abroad, 1977*, p. 16.

13. Ibid., p. 17.

14. P. Stoneman, "Patenting Activity: A Re-evaluation of the Influence of Demand Pressures," *Journal of Industrial Economics*, June 1979, pp. 385-401.

15. R. W. Wilson, "The Effect of Technological Environment and Product Rivalry on R & D Effort and Licensing of Inventions," *Review of Economics and Statistics*, May 1977, pp. 171-78; W. H. Davidson and D. G. McFetridge, "International Technology Transactions and the Theory of the Firm," *Journal of Industrial Economics*, 32:253-64, March 1984.

16. J. H. Dunning, "Non-Equity Forms of Foreign Economic Involvement, and the Theory of International Production," working paper, University of Reading, 1982; S. Kobrin, "The Environmental Determinants of Foreign Direct Manufacturing Investment: An Ex-Post Empirical Analysis," *Journal of International Business Studies*, Summer 1976, pp. 29-42.

17. Casson, *Alternatives to the Multinational Enterprise*; P. J. Buckley and H. Davies, "The Place of Licensing in the Theory and Practice of Foreign Operations," University of Reading Discussion Paper no. 47, November 1979.

18. Contractor, *International Technology Licensing.*

19. W. H. Davidson, "Location of Foreign Direct Investment Activity: Country Characteristics and Experience Effects," *Journal of International Business Studies*, Fall 1980, pp. 9-22; Kobrin, "Environmental Determinants"; R. T. Green and W. H. Cunningham, "The Determinants of U.S. Foreign Investment: An Empirical Examination," *Management International Review*, vol. 2-3 (June 1975), pp. 113-20.

20. Casson, *Alternatives to the Multinational Enterprise*; Buckley and Davies, "Place of Licensing."

21. F. R. Root and A. A. Ahmed, "Empirical Determinants of Manufacturing Direct Foreign Investment in Developing Countries," *Economic Development and Cultural Change*, vol. 27 (July 1979), pp. 51-67.

22. Contractor, *International Technology Licensing.*

23. Ibid.; Dunning, "Foreign Economic Involvement."

24. Davidson, "Location of Foreign Direct Investment."

25. For details on its construction, see Business International, *Managing and Evaluating Country Risk* (New York: Business International Corporation, 1981).

26. Kobrin, "Environmental Determinants"; Contractor, *International Technology Licensing.*

27. Casson, *Alternatives to the Multinational Enterprise.*

28. Ibid.; S. P. Magee, "Technology and the Appropriability Theory of the Multinational Corporation," in *The New International Economic Order: The North-South Debate*, ed. J. Bhagwati (Cambridge, Mass.: MIT Press, 1977).

REFERENCES

R. Z. Aliber. "A Theory of Direct Foreign Investment." In *The International Corporation*, ed. C. P. Kindleberger. Cambridge, Mass.: MIT Press, 1970.

P. J. Buckley and H. Davies. "The Place of Licensing in the Theory and Practice of Foreign Operations." University of Reading Discussion Paper no. 47, November 1979.

Business International. *Managing and Evaluating Country Risk*. New York: Business International Corporation, 1981.

M. Casson. *Alternatives to the Multinational Enterprise*. New York: Holmes and Meier, 1979.

R. Caves. "International Corporations: The Industrial Economics of Foreign Investment." *Economica* (February 1971), pp. 1-27.

F. Contractor. *International Technology Licensing: Compensations, Costs and Negotiations*. Lexington, Mass.: D. C. Heath, 1981.

W. H. Davidson. "Location of Foreign Direct Investment Activity: Country Characteristics and Experience Effects." *Journal of International Business Studies* (Fall 1980).

W. H. Davidson and D. G. McFetridge. "International Technology Transactions and the Theory of the Firm." *Journal of Industrial Economics* 32:253-64, March 1984.

J. H. Dunning. "Non-Equity Forms of Foreign Economic Involvement, and the Theory of International Production." Working paper. University of Reading, 1982.

———. "Towards an Eclectic Theory of International Production: Some Empirical Tests." *Journal of International Business Studies* (Spring 1980).

M. Finnegan and R. McCarthy. "U.S. Tax Considerations in International Technology Transfers." *Licensing Law and Business Report*, April and May 1979.

B. Gottfried and J. Hoby. "Wirtschaftspolitik gegenüber Auslandscapital." *Bulletin of the Sociological Institute of the University of Zurich*, no. 35 (1978).

R. T. Green and W. H. Cunningham. "The Determinants of U.S. Foreign Investment: An Empirical Examination." *Management International Review*, vol. 2-3 (1975), pp. 113-20.

S. Hymer. *The International Operations of National Firms: A Study of Direct Investment*. Ph.D. dissertation, Massachusetts Institute of Technology, 1960.

S. Kobrin. "The Environmental Determinants of Foreign Direct Manufacturing Investment: An Ex-Post Empirical Analysis." *Journal of International Business Studies* (Summer 1976).

G. Kopits. "Intra-Firm Royalties Crossing Frontiers and Transfer-Pricing Behavior." *Economic Journal* (December, 1976).

S. P. Magee. "Technology and the Appropriability Theory of the Multinational Corporation." In *The New International Economic Order: The North-South Debate*, ed. J. Bhagwati. Cambridge, Mass.: MIT Press, 1977.

R. D. Robinson. *National Control of Foreign Business Entry*. New York: Praeger Publishers, 1976.

F. R. Root and A. A. Ahmed. "Empirical Determinants of Manufacturing Direct Foreign Investment in Developing Countries." *Economic Development and Cultural Change* (July 1979), vol. 27, pp. 51-67.

M. Sato and R. Bird. "International Aspects of the Taxation of Corporations and Scholars." IMF Staff Papers, no. 22, July 1975.

P. Stoneman. "Patenting Activity: A Re-evaluation of the Influence of Demand Pressures." *Journal of Industrial Economics* (June 1979).

D. Teece. "The Market for Knowhow and the Efficient International Transfer of Technology." *Annals of the Academy of Political and Social Sciences* (November 1981).

——. *The Multinational Corporation and the Resource Cost of International Technology Transfer*. Cambridge, Mass.: Ballinger, 1977.

U.S. Department of Commerce. *U.S. Direct Investment Abroad, 1977*. Washington, D.C.: Government Printing Office, 1981.

R. Vernon. "International Investment and International Trade in the Product Cycle." *Quarterly Journal of Economics* (May 1966).

R. W. Wilson. "The Effect of Technological Environment and Product Rivalry on R & D Effort and Licensing of Inventions." *Review of Economics and Statistics* (May 1977).

Index

About the Editors

Nathan Rosenberg is Professor of Economics (and currently chairman of the Economics Department) at Stanford University. He has written extensively on the economics of technological change. His most recent book is *Inside the Black Box* (Cambridge University Press, 1982).

Claudio Frischtak is an industrial economist at the World Bank. He will soon complete a doctoral dissertation at Stanford University dealing with the role of technological change in Brazilian exports.